AF696335

Copyright © 2018 Irwin Ki All rights reserved.
No part of this publication may be reproduced, distributed, or transmitted in any form or by any means, including photocopying, recording, or other electronic or mechanical methods, without the prior written permission of the publisher, except in the case of brief quotations embodied in reviews and certain other non-commercial uses permitted by copyright law.

ISBN-13: 978-1-7328953-0-0

Please note: some names and identifying details have been changed to protect the privacy of individuals featured in this book. However, the names and identifying details of interviewees were not changed. The interviews themselves were edited in places simply to make them more readily understandable.

Whether you end up in technology or not, this book is dedicated to the young and restless, the ones who are looking to make a difference in the world. Despite the news and media selling fear on machines and artificial intelligence replacing human jobs, I'm an optimist on how we can use technology for the better of humanity. This book is my contribution so that even more people can join in helping build a great future for everyone, and it starts with getting in the game.

ALL the stories in this book are true. While a few names were changed for privacy reasons, the information you will find in the interviews is tried, true, and proven to be successful.

This book is dedicated to the next generation of leaders and change-makers.

CONTENT

SECTION ONE

WHY TECH?

Hi! I'm Irwrin!

CHAPTER 1

FAILING INTO SOFTWARE ...

Hi, I'm Irwin, the author of *Breaking into YOUR Silicon Valley*.

I've made it my mission to solve problems in the tech sector. I've helped IT buyers evaluating and procure enterprise software, and I've help build sales teams and net new business for sales leaders and the entrepreneurs and investor they support. In this book I'm sharing **INSIDER SECRETS** and compiled **TRIBAL KNOWLEDGE** on how I and many others like you got their start in tech—no matter their prior experience or where they are in the world.

I don't have a computer science degree. I didn't spend years in my basement writing code for the fun of it. As a matter of fact, I still don't know how to write code! Yet, I'm well into a successful and fulfilling tech career.

I've worked at big tech giants—Cisco and BMC Software—and I've worked with other entrepreneurs in smaller tech startups before they've become a well know brand name within the industry—OKTA (IPO'ed) and OpenDNS (acquired by Cisco), a subsidiary with around 300 employees underneath Cisco's giant 70,000+ employee organization.

So, while I may not be a computer whiz, I managed to "crack the code" of breaking into a tech career as a non-techie regardless of where you are located. And, in this book, that's what I'm going to teach you how to do too.

Let's start first with WHY I'm writing this book. Silicon Valley has been in the center of a lot of controversial news lately. There is a lot of attention that automation, artificial intelligence, and disruption is going to displace A LOT of jobs in the future. There is fear that big technology companies have too much power and control over our data and privacy. There's even been a book (*Brotopia by Emily Chang*) written about the struggles the tech industry is having with diversity and treating women fairly. My belief is the tech sector has major improvement ahead and it is nowhere near perfect. I don't want the negative news to outshine on the HUGE opportunity that is available around the globe.

Notice that this book's title is "Breaking into YOUR Silicon Valley" vs. "Breaking into Silicon Valley." There are going to be massive amounts of improvements in healthcare, finance, transportation, and it won't be limited to Northern California. I got my start in North Carolina and with technology popping up everywhere, there is an opportunity for those willing to reach out and ride the wave.

I believe that to solve many of today's challenges, we will need talent and voices from different backgrounds, perspectives, and genders. That is what I hope this book will do. I hope that the stories shared, and the lessons I've personally walked will inspire you, or someone you know, to believe that you are more than capable of joining this movement.

My WHY is to help provide transparency, teach, and ultimately inform people's mindsets so that there will be a flood of top-of-the-funnel hiring so new voices from different parts of the globe can position themselves to be part of this movement. I see three main readers of this book:

1. First are the college graduates (or soon-to-be-graduates) who still have no idea what to do or need help in taking their degrees and turning them into opportunities.
2. Second are the employees who are at dead-end jobs (or coming back into the workforce) and feel that they are just a cog in the machine and are likely in an industry that will be disrupted by technology.
3. Third are the aspiring tech entrepreneurs who want to get in the game but are not sure of the different skill sets they need to succeed.

Here is what is possible.

For the college graduates or those who feel stuck and have no idea what they want to do, this book will allow you to sponge up decades of experience across many different segments of business. You'll be able to get clarity on where you want to direct your life. I've found that the biggest challenge is getting inspiration on where to focus and who you ultimately want to become, especially when you are just starting out.

By jumping into tech via the method described in this book you'll be part of a movement that is shaping the course of history as we speak. Plus, you'll get paid to develop high-income life skills along the way. You will be part of the tech community. The people you'll meet and become friends with (maybe even your future spouse) are some of the smartest and brightest in the world. They consist of dreamers, risk-takers, doers, investors, and even celebrities (google this phrase: "top celebrity startup investors").

Even if your first role in tech isn't so glamorous, at least you are in the game. You'll see where to go next or whom to meet. As your career progresses, you'll be bombarded with recruiters who will pitch you in joining the next big thing.

For readers who are aspiring entrepreneurs, you could meet your co-founders and early hires being in the game. I've seen my personal friend find his co-founder by bumping into him at the Google cafeteria. One of my other friends found the courage to leave and start his own business after learning of a pain point in the market that wouldn't have been apparent if he were outside of tech. The name of the game is to get IN the game!

For those skeptics out there, who are thinking along the lines of, "Sure, Irwin, you might not be able to code, but you're probably a hardcore mathematician" or "Irwin, you probably have a best buddy who just gave you a job, but I'm not connected like that,"give me a moment to fill you in.

GOLDEN NUGGET

For those who are wondering if I'm too old or have been out of the workforce too long. I've found that the antidote for those concerns is to look at this book and the lessons with eyes of your younger self. What I mean is looking to look at this book as a fresh opportunity to grow, give, and be a service to others. Yes, it'll might require the "hustle" but what you may lack in youth you make up with wisdom and experience. Which is much needed. Look at your weakness as a strength...my only piece of advice is to focus on your health so you can run with those younger than you. There are folks in the following chapters that are older than the age of 40 who still have the energy and enthusiasm of when they first started.

Are you familiar with

» I'm about to graduate from college, and I don't know what I'm going to do next. *Feeling anxious and confused.*

» I'm working at .... because I thought I'd love it, but it's not at all how I imagined. It kind of sucks. *Feeling stuck and bored and anxious.*

» My parents, friends, family, girl/boyfriend are pressuring me to explain my plan for the future, but I don't have one. *Feeling the pressure.*

» I wish I could be a part of the excitement in Silicon Valley, in tech, but I'm not a computer science geek. Computer languages? Just speaking English correctly is hard enough. *Feeling unlucky.*
» I'm not near Silicon Valley, and I want to stay near my family (Addressed in Chapter 15). *Feeling stuck*

Me too. Here's my story.

After a month at NC State I knew deep down that I'd picked the wrong major: civil engineering—you know, doing stuff with bridges, highways, and surveys. I felt trapped and stuck (barely passing with a GPA of 2.1). But, my parents and mentors advised me that an engineering degree would "hold up better" than a business degree.

As the months turned into semesters, I knew I had to figure out what I wanted to do. My dissatisfaction with my major, combined with the clock counting down to graduation, led me to accelerate my exploration of jobs, internships, and life experiences, all in an attempt at finding the career path that was right for me.

Real estate agent, civil engineer, loan officer, entrepreneur—I tried on all these hats. Some I wore for longer than others. I even went overseas to see how a CEO lived. I moved to the Canary Islands and worked as a day trader and poker player.

By my final semester of college—I'd made it ... nowhere. I'd been fired from two different places: Audi and a local digital agency. The real estate flipping company I'd started went under. Two of my business plans went unused. And the cherry on top: it was 2010, just two years after the great financial crisis of 2008.

Time was up.

And even with my beloved engineering degree—you know, the one that would "hold up" above all others—I couldn't get hired.

After pouring out my struggles to a friend, he suggested I look into technology.

My first response: "I have no tech background. I have no computer science degree." Frankly, I'd never even looked at that world because it just seemed too complex. Too technical, for me at least.

His response: "You know, Irwin, there are a lot of entry-level roles that don't require any prior tech experience. You'll get on-the-job training. If you have the drive, the hunger to learn, and the humility to take coaching, you have shot!"

I was shocked.

TIME IS A CRUEL MASTER.
NO, TIME IS A THIEF, AND A VILLAIN.

— ALICE'S ADVENTURES IN WONDERLAND

It took me a total of five months, but with a lot of persistence, trial and error, and caffeine, I finally broke into tech! I managed to get my foot in the door, and I was determined to see where this journey would lead me.

What was that tech job I landed? I started as a sales development representative at BMC Software. And that was the start of my career in tech sales. I was tasked with calling into the largest West Coast companies—Disney, PayPal/Ebay, 20th Century Fox, MGM Resorts—and I was paired with the enterprise account executives who were tasked with selling huge software contracts to the IT buyers in the largest companies in the world. I'm talking multimillion-dollar deals.

In my first week, when I sat down with my manager, he asked, "Do you realize the opportunity at hand here? I was at dinner with the top four reps from the previous year. Every single one of them had made over $1 million in commission."

WOW! That's when I realized—*there's a huge opportunity here—in tech—for me.* It was at BMC that I met Adam Aarons who became my mentor. Adam eventually left BMC to join a San Francisco startup called OKTA. He and the executive team then hired me, at 24 years old, to build and manage the newly-created sales development representative (SDR) team. After much discussion with my girlfriend (then turned fiancée), we moved our entire lives across the country, from North Carolina to San Francisco, CA.

Over my first year I hired many SDRs from very different backgrounds. One had worked at Macy's, another at a Sprint store, another selling persian rugs, someone had been at a radio station, another had worked at Enterprise Rent-A-Car, as well as a lot of recent graduates. I hired history, business, and English majors. (Notice: these hires didn't have CS, math, or tech backgrounds.)

That's when I had my first of many breakthroughs: **SDR is a great entry-level position** into the technology business. There is no correlation between college majors or gender when it comes to being a successful SDR. If a person is hungry and willing to learn, work hard, and align to the corporate culture—that's it. They'll only climb higher in their sector of tech.

When my wife, who hadn't graduated from college, went from working the front desk at a chiropractor's office to becoming a recruiter at a small company in Menlo Park called Facebook :) —I'll be giving you the details of her successful break into tech later in the book—that's when I had my second insight: **there are entry-level jobs (what I call gateway jobs) in every department in every tech company.**

Gateway Job = Entry-Level Job + High Growth Company

At first glance it seems humble. The difference isgateway jobs leads to many different career paths across different companies and can happen quite fast if you are at the right industry!

And again, as my wife showed if a person is hungry and willing to learn, put in the work, and align to the corporate culture—that's it. They'll only climb higher in their sector of tech.

Then I had an ah-ha moment! I realized that every job segment had to have a gateway role. For me it was my first job as a sales development rep. From my wife it was first getting experience as an office manager, then the gateway job was a recruiting coordinator position. So I took some time and mapped out what are the different segments of business. I wrote down sales, marketing, operations, finance, product management, developer, people (HR), customer support, IT, and legal. Then I asked myself, "What are there other gateway roles for each of the business segments?"

Then I realized there was nothing like this on the internet. **It was all tribal knowledge.** That's when I became obsessed with decoding every "gateway" job into tech.

I realized that I only knew 2 gateway roles and I need help. So I reached out to numerous friends, colleagues, and folks who worked at companies like Facebook, Salesforce, Cisco, OKTA, and many more. Their response and willingness to offer their voice was amazing! One by one we recorded an informational interview (more on Chapter 17) and asked each of them to answer three fundamental questions:

1. What's your story and how did you break into tech?
2. What are the gateway job(s) for your area of tech?
3. What is your advice for others looking to break in?

This is only one part of what this book is going to deliver: an inside look from 10 different gateways each with their own insider sharing their story, knowledge, insight, and advice on how to make it in their segment (e.g., sales, marketing, HR, support, finance, legal). Here's the crazy part—after doing over a dozen interviews, I've found that one out of four folks broke into technology without a college degree—something I find so different than what I've seen in other industries.

I've come to realize there is a complete lack of transparency on the different types of roles within any business organization. We were simply never taught what types of careers are possible and the clear entryways to break in. The "gateways" portion of this book (section 2) gives my peers' individual stories, so you can gain a good idea of what a career in sales or customer success looks like. It should give you a shortcut to figuring out what part of the business you can see yourself in, which should save you a TON of time.

Let's assume you know what you want to do regardless of tech or not. Part of the challenge is not knowing what to do next. Too many times we are simply expected to figure it out ourselves. This is why I'm pumped that section 3 of this book, The How—Putting Words to Action, where I have tirelessly documented the ENTIRE process of helping out two lucky mentees go from blank resume to negotiating multiple-offer scenarios. Also, I've included the many lessons I realized I took for granted and new lessons I've learned doing the entire process from beginning to end.

Every semester students are graduating in droves, yet never once did they have a class on how to figure out their strengths and to match those strengths to gateway roles. They likely never got advice on how to write their resumes and LinkedIn profiles, how to find and build a list of target companies, and how to break into the career of their dreams.

Another insight I got was how these gateway roles are prevalent in most big cities! Yes, Silicon Valley is the most famous. But remember, I got started out in North Carolina. You'll read an interview on how my friend Lenn started at a gateway within Fidelity Investments before moving out to San Francisco. The point I'm trying to make is you can create your own Silicon Valley journey.

In this book I have also included career advice from Silicon Valley and technology's finest: Reid Hoffman, Mark Cuban, Jason Calacanis, Eric Schmidt, and others. I've taken the lessons I've learned from selling to the largest enterprise, being on the hiring side of the table, and working at two startups that had successful exits. You'll find a unique perspective on how to pick a successful startup and insider secrets to help you break into an industry that is actively changing and disrupting the world. This book is a compilation of what I would do if I had to do it all over again when breaking into and creating a successful career in tech.

The first few things you need to do: believe you are worth it and invest in your great worth by reading this book. And if feel someone in your life, a friend, an acquaintance, or family member, needs clarity and direction please, share this book with them.

I've repeatedly been saying that a tech career is amazing because it offers fulfillment, success, prosperity, and all that other good stuff. Chapter 2 will explore this more in order to convince you that tech is the place for you. Also, chapter 2 reviews how this book is laid out so that you can be sure to get the most out of it.

CHAPTER 2

THE "T" WORD

FACTS

- Half of employed U.S. college grads are in jobs that require less than four-year degree
- There is soooo much student loan debt. 4 out of 10 borrowers are not making payments.
- The average American saves less than 5% of their income.
- Not everyone is meant to be an entrepreneur (at least immediately).

How do we turn these facts to myths?

With the "T" word.

It's time for me to let you in on my secret ingredient for finding your next step in life. Don't get scared. Don't put this book down. Don't think you've wasted your money.

My secret ingredient for getting what you want out of life and finding the career you want is ...

Technology!

I know what you might be thinking:

> "Irwin, I'm not good with technology."
> "Irwin, I don't have a degree that has anything to do with technology."
> "Irwin, do you know how smart you have to be to have success in the technology field?!"
> "Irwin, I'm not passionate about technology at all. I'm an old-school paper-and-pencil kind of person. I want to be able to do what I love!"

Perhaps you truly believe these things about yourself, but just give me a chance to change your mind. Actually, let me put it this way: just give me a chance to open your mind. You see, those thoughts and fears are all misconceptions and limiting beliefs.

Trust me. A career in technology isn't what you think.

In the non-tech world it could take years before you get a promotion or even worse you are stuck. Think front desk at a doctor's office vs. front desk at Instagram. Most regular (non-tech) companies grow slowly, so your time at them also seems to go slowly. Aside from a corporation's stock purchase plan, it's very unlikely a business owner will give you a piece of their business. Your role is likely fixed with very little challenge or growth that happens after the first few months.

Technology startups and corporations offer some of the best ways to accelerate your skills, finances, and network. Instead of hoping and waiting over the years for a promotion and greater responsibility, in tech it can happen in just months. You are paid to learn and grow, and the life experience gained can position you to become even more sought-after in your blooming tech career. It's usually fast-paced, and you are often asked to take on more, and that's a good thing for your personal growth.

And that computer science degree you think everybody in tech has? Not the case. Not at all. Yes, part of the team have CS degrees, but the reality is that most come from other backgrounds—from history to studio art to biology to no college degree at all! I bet you are 100% stunned by this tech reality!

Why Tech Is the Place for YOU

❶ Growth and Innovation

It's no secret; you can't escape technology. It is the future. I mean, just look at all the advances we have made in the last 10 years alone in cell phones, watches, healthcare, transportation, and currency! I can literally ask my watch a question, and it will give me the answer. Crazy! Everything is connected, and every company is using technology (and if the company isn't, then it's going to die—just look at what happened to Blockbuster and Polaroid. THEY ARE GONE!).

There is also huge potential for personal growth as well. There are always new startups, new initiatives, new goals, and new offices opening, which will require more hiring, which will require more leadership, which means a lot of times you are going to be stretched! You'll be challenged, and with challenge comes the opportunity to shine. With fast growth comes new leaders and executives needed to lead the realm. That is your opportunity.

❷ Upside

When Mark Cuban was asked, "What would you recommend for someone going into the workplace?" he suggested, "If you can take a job that allows you to earn equity in your company, that's always a huge advantage. It doesn't have to be a tech company. It can be a company that makes bricks. If you have the opportunity to work for a company that offers you equity, that offers you a chance to participate in the upside, particularly if it's early in its life cycle, that's a chance to really do well." This is one of my favorite reasons why technology is a very interesting space. If you think about it, in the past century, there hasn't been a lot of companies that actually gave part ownership (through stock options, aka equity) to their employees. Even today, small business owners rarely give their employees a piece of the upside. Most companies have income and some sort of health or retirement benefits, e.g., 401(k)s. The technology sector is the only one where I've seen ALL employees, both from entry-level all the way to C-level executives, get offered a piece of the upside (equity).

❸ Competition

Competition is GOOD for employees—and those seeking to be employees—in tech. With so much disruption and innovation, there is a lot of money to be made. There's plenty of investors looking to find entrepreneurs to fund. But there lies the problem. The existing established companies, like Google, Facebook, and Apple, have the brands, big budgets, and competitive pay. They are constantly stealing and battling over talent. Talent (i.e., you) has become special. You are the prize. There is competition to get you (yes, even if you are an English major or college dropout)!

❹ An Insider's Peek into the Future!

We're living in a world where we're seeing self-driving cars get put on the road in front of us. We're seeing cybersecurity companies get created to protect us against the next generation of threats. We are seeing how cryptocurrency is disrupting finance, and how the future of money is getting transformed.

Because our future is being created and conceived right now, if you decide to involve yourself in tech, you have the opportunity to make bets 1 to 1,000 years ahead of the pack. Imagine if you could see the next ten years' stock prices. It's kind of like that. By immersing yourself in a tech career, you uniquely position yourself to have an inside view on the future. You can leverage your special tech access to make investments that others would never consider.

THE SECRET TO SUCCESSFUL HIRING IS THIS: LOOK FOR THE PEOPLE WHO WANT TO CHANGE THE WORLD.

— MARC BENIOFF, CEO OF SALESFORCE

5 Impact

It's time to bring in the millennial billionaire himself, Mark Zuckerberg. He's impacting the world in ways that we don't even realize: American and global politics, bringing together families and friends from across the globe, educating people in one place about the environmental and political realities of those in another. Currently, the world is watching him and his team address the future of data privacy.

What I love about technology is it's pervasive in everyday life. Everyone has a powerful computer in their hands where they can learn programming, check out the news, communicate with family, or go to the darkest, deepest parts of the web. It's ubiquitous and expansive. This wasn't possible a decade or even two decades ago. By working in tech, you position yourself to make huge impacts on individual lives.

6 No Experience or CS Degree Required

I broke into software with no prior technology experience. My college degree was in civil engineering, meaning I was prepped to help out in the construction of bridges and roads or in doing land surveys. What this means is that I was not at all a so-called "techie." However, with persistence (and lots of caffeine) I broke into the tech world and worked my way up! I've been a top software sales rep across four different companies. I've helped entrepreneurs successfully exit two startups, OKTA (IPO) and OpenDNS (acquired by Cisco), meaning that after years of being funded by venture capital and running in the red, the investors' shares became liquid through an acquisition or the open public could invest in the shares on the open market.

IT IS POSSIBLE.

The Irwin Ki Guarantee: I'm committed to developing the future leaders of technology and software. My goal is to share the wisdom I gained from my own personal battles as well as the lessons imparted to me by my mentors, peers, colleagues, and tech hires. *Breaking into YOUR Silicon Valley* will shorten your learning curve and show you the openings so that you can break into tech as quickly—and as lucratively—as possible.

Breaking into YOUR Silicon Valley will supply you the jumpstart on the competition. I'm going to give you a heads-up on the often unconsidered gateways into tech careers as well as the roadblocks and blind spots you may encounter along the way. You will learn how exciting and abundant a career in software can be.

Getting the Most out of This Book

This book is broken up into three sections:

You are reading the first part on why technology is such a hot space.

The second part helps to figure out which "gateway" into a tech career makes the most sense for you. Think of a "gateway" as the entry point through which you can start your rock-star tech career. A big mistake a lot of people make is thinking that there's only one gateway into tech: being a genius coder with a CS degree. In reality there are multiple gateways for breaking in. In *Breaking into YOUR Silicon Valley*, we'll explore the following:

» The Sales Gateway
» The Marketing Gateway
» The Operations and Analyst Gateway
» The Finance Gateway
» The Product Manager Gateway
» The Developer Gateway
» The People (HR) and Recruiting Gateway
» The Customer Support Gateway
» The IT Gateway
» The Legal Gateway

Each gateway offers a lifetime of fascinating impactful work with great people and large salaries. It's important not to get thrown off by the salary numbers next to each gateway. I want to EMPHASIZE that these gateways more often than not lead to six-figure compensation plans. Remember, it's a start, and a lot of times you are paid to learn.

To determine your ideal gateway into an interesting and prosperous career in tech, for each gateway, you will encounter:

- The particular tech career paths it can open to you
- Compensation plans i.e., $$$. **(Please note: all salary info is pulled from Glassdoor.com and does NOT include benefits, bonuses, commission, or equity)**
- Duties and responsibilities
- Required skills, university degrees (if any), and personal characteristics
- Analyses of real job postings from tech companies
- Stories about and interviews with real people in the tech community pursuing the given gateway
- And more!

This will show you the whole road ahead, the lessons learned, and, most importantly, the various places you can start your career in tech.

Half of the book is the gateway section. Even though it seems very long, it's designed to be skipped ahead! Just an FYI, the interview sections in the book have a TON of wisdom and stories. Once you determine your ideal gateway (feel free to just read the gateways that you care about), it's time to take smart action.

Part 3 is the meat of the book. It contains insider secrets and proven strategies on how to land that first technology job anywhere in the world.

In part 3 you will learn to:

- Set up an action plan using a Mutual Action Plan (MAP) to land your chosen gateway job.
- Consider and determine location options that work best for you
- Decide the kind of tech company you want to work for: seed, early- to mid-, or large and established
- Craft an effective resume and LinkedIn profile, particularly if you have no experience
- Apply online and get recruiters to find YOU!
- Land referrals from insiders
- Slam dunk interviews: the dos and don'ts
- Negotiate to get the best offer possible

Oh wait! That's not all. I'm also including two bonus chapters: the first is on breaking into cryptocurrency and the second addresses breaking into the gaming industry. You'll find a load of insights and insider info for both these different but exciting industries.

As you can see, Breaking into YOUR Silicon Valley is the comprehensive, insider's guide to breaking into tech. This book is going to give you a slew of information from me, as well as others. Our careers are testaments to what we teach, preach, and reveal in these pages. It's time we explore those gateways, beginning with the one that launched my career—the sales gateway.

GET YOUR FREE BONUS!

Thank you for reading & sharing this book!

To help with your **momentum!** Here are 4 **FREE videos**

I've included for you to bring the **book to LIFE!**

Go To: www.irwinki.com/p/FREETRAINING

If you want to see more videos!

Follow me: Search "Irwin Ki" on Youtube

Find me via @Irwin_Ki on Instagram

Get **connected with other readers** via FB group:

"Breaking Into Your Silicon Valley Community"

SECTION TWO

THE GATEWAYS

CHAPTER 3

THE SALES GATEWAY

We are going to begin with one of my favorite gateways: sales. This gateway is where most executives—like Mark Cuban, serial entrepreneur, investor (Shark Tank), and owner of the Dallas Mavericks—started their careers. Every tech company is selling something—whether it's a product, like Apple or Samsung, or it's your attention and data, like Facebook and Google. Maybe it's time savings, like what Uber or Amazon sells. Either way, there are sales happening!

Regardless of which gateway you choose, at the end of the day, *you* are selling yourself. Another reason why sales is a powerful gateway is because it's applicable to other careers. I have seen many people break into other fields—marketing, recruiting, business development, solutions engineering, operations, etc.,—when they leverage the sales gateway.

Why? Because it solves the #1 challenge tech companies have with hiring people from non-tech backgrounds. The sales gateway allows you, starting from zero experience, to learn three main things, so you can move to the other fields. Those three things—you learn how to talk about:

1. The product
2. The market
3. The customers you are serving: their needs, wants, desires, and how to serve them

Career Paths:

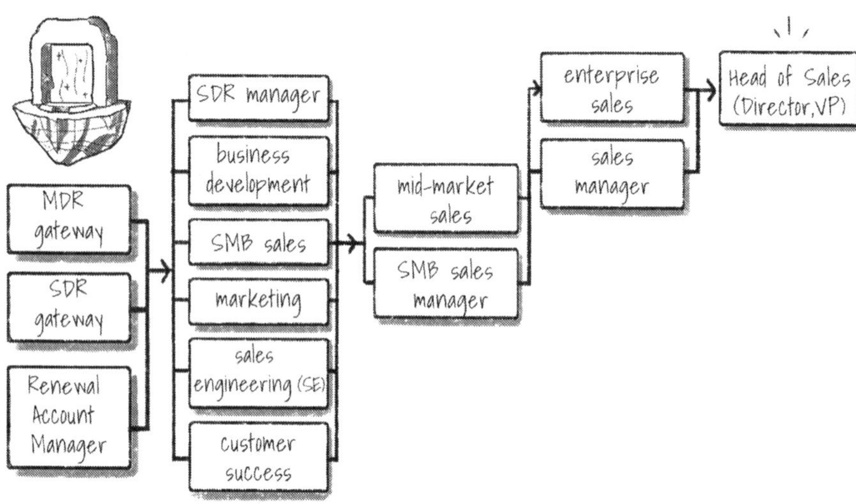

Gateways

» Sales development representative, SDR ($64k–$91k)
  - Also known as "business development rep," BDR
  - Sometimes also called "marketing development rep," MDR
» Renewals account manager ($80k–$130k)

Main Sales Career Paths

- » Account executive (AE)—help find new customers to BUY products or services ($100k–$1M)
- » Account management (AM)/renewals—help keep those customers happy and spend more over time ($100k–$200k)
- » Business development (BM)/partnerships/channels—help find other companies that will mutually benefit from sharing their resources to ultimately lead to more people buying products or services ($100k–$250k)

Regardless of the stage of a business, every company has a need to grow revenue. For business to consumer companies (think Apple and Uber), the majority of sales is controlled by leveraging marketing. For business-to-business (B2B) sales, there is a direct sales team that drives sales, and a huge organization that supports it. I'm not going to cover leadership roles, but rather we'll dive deep into the individual contributor roles that help support a sales organization.

The challenge of being great in a sales role at a tech company for people from a non-tech background lies in starting from not knowing anything. In the sales development role, you are trained on three absolutely critical items to add value to any organization:

1. Talking about the product
2. Learning about the market.
3. Learning about your customers

Once you master these three areas, all the while working diligently and building internal champions (see chapter 9 for more on that), then you can succeed in almost all other tech companies as well as many other career paths in tech itself.

Sample SDR Career Description

Don't believe that you can start with zero experience? I've pulled an SDR career description. Notice the parts in **bold**; these are the requirements the company is looking for. Notice how these requirements allow people with zero experience the possibility of getting hired:

> Sales development representative is an excellent opportunity to help drive demand for our market-leading cloud adoption and management solution. Your consistent track record of working strategically with both marketing and sales teams in a business development and lead generation capacity will allow you to thrive in our dynamic, high-activity environment.
>
> Working closely with marketing, you will own reaching out to prospective customers and for qualifying opportunities. Working with our seasoned account executives, you will map out and build pipeline, and help resolve our most effective marketing channels. We bolster promoting from within, and with our current growth plans and the acceleration we're seeing in our business, strong performance in this role can lead to upward mobility within the sales organization.
>
> Our sales representatives are **high-energy, self-motivated, agile,** and able to change hats on the fly. In this highly prized role, **excellent phone and email skills, and the ability to quickly assess customer opportunities** will be critical to your success. You will be responsible

for accurately tracking customer interactions and information in our CRM system.

We're looking for professionals with **strong organizational and follow-up skills who work well in a team environment.** Prior lead generation or sales prospecting experience and a college degree will enhance your candidacy for this role **[But notice that they aren't required].**

If this sounds like the type of environment in which you thrive, then we want to talk to you about this career-defining opportunity.

Job Duties and Responsibilities

» Generate new business opportunities to fuel the sales pipeline
» Efficiently respond and qualify inbound marketing leads
» Help build and prioritize strategic target account lists within a defined territory
» Research and build new and existing accounts (i.e., adding contacts, emails, strategic calling)
» Conduct high-level conversations with senior executives in prospect accounts
» Evangelize!
» Achieve monthly quotas of qualified opportunities and closed business

Required Skills

» **Minimum of** 0–2 years of previous prospecting experience
» Salesforce.com experience
» Excellent written/verbal communication skills

I HAVE THE CAREER ADVICE WHETHER YOU ARE IN COLLEGE, GRADUATING, OR HAVE BEEN IN THE WORKFORCE FOR 3–5 YEARS. GET SALES EXPERIENCE. SALES EXPERIENCE WILL HELP YOU OUT NO MATTER WHAT YOU DO IN YOUR CAREER.

— MARK SUSTER'S SNAPCHAT: @MSUSTER
SENIOR PARTNER UPFRONT VENTURES

» Strong analytical skills
» Highly motivated, driven, and self-starting
» Ability to work in a dynamic team environment
» Ability to understand customer needs and meet those needs with a successful product sale
» Excellent time management/organizational skills
» Bachelor's degree from an accredited university (debatable)

To help paint the picture of how flexible the SDR gateway is, here is a short story of four SDRs who started together and the different career paths they choose after 4+ years. This will give you an idea of what the skills and degrees each person had before getting hired as well as the other tech careers paths they landed after they successfully entered the sales gateway.

My Vision for You

Most people growing up never dreamed of doing sales for a career. In fact, most cringe when they think of sales. If you can see the bigger picture of it being a stepping stone to the many career paths in technology, then by investing in mastering the SDR game for 1 year, you will have the experience needed to choose your desired path. Don't believe me? Here are four folks I personally hired, mentored, and saw succeed!

Here are the four people we will be discussing:

Alex, Parker, Karina, and Kevin were part of the original SDR I hired at OKTA. They all have very unique backgrounds.

These four people went through onboarding, learned from the SDR manager, and worked with leaders within the company to learn the ins and outs, the technical and business use-cases.

Two weeks from their start date, they were mapped to region-specific sales teams. Their specific tasks involved reaching out and qualifying current and old marketing leads, running outbound campaigns through email, doing call and voicemail reach-outs, and booking meetings for the seasoned sales teams.

In the process, they were exposed to different business units including marketing, renewals, business development. The marketing manager wanted to ensure that the leads were prioritized and were called throughout the week consistently. They sometimes dealt with support when customers had questions, and it was necessary to open up a technical case. They also joined account managers, customer and forecast calls, and learned how a sales process was run from the first call to what the AMs needed to close their deals. So, by working in sales and starting with little experience, these four were exposed to most areas of the tech company as a whole.

Kevin: Route to SMB > Mid-Market > Enterprise Sales

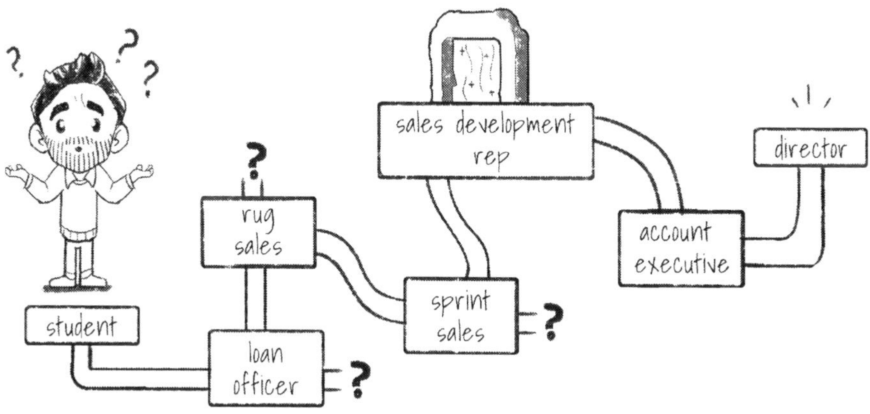

Kevin is the oldest of the bunch. He immigrated to the United States at 24 from Iran, and after graduating from USC, he was a loan officer before the real estate financial crisis of 2006–2008. Next, he went on to selling antique rugs, and when he applied at OKTA, he was slinging cell phone plans at the local San Francisco Sprint store.

Right out of the gate, Kevin worked the hardest. He came in early and left late. His numbers reflected his persistence too! He was the #1 seller and the most professional and mature. It was unanimous that he was the top pick for getting promoted to the small and medium business (SMB) side of OKTA after only six months of being an SDR.

After another year, he again was the top of the SMB team, so again he was promoted. Kevin moved to Georgia and built the southeast market for mid-market-sized (corporate) companies. He continued to be a leader and top performer, and moved on to do enterprise sales where he was selling to the largest customers.

Kevin is a great example of what it takes for a person to move to a field role in the fastest and most focused track possible. The account executive (AE) role requires a desire to win, along with the polish of a business professional, the listening and questioning skills of a consultant, the intelligence to navigate complex deals, the willingness to deal with failure, and an awareness on how to manage up to his leaders!

Alex: Route to Business Development, Channels, and Partnerships

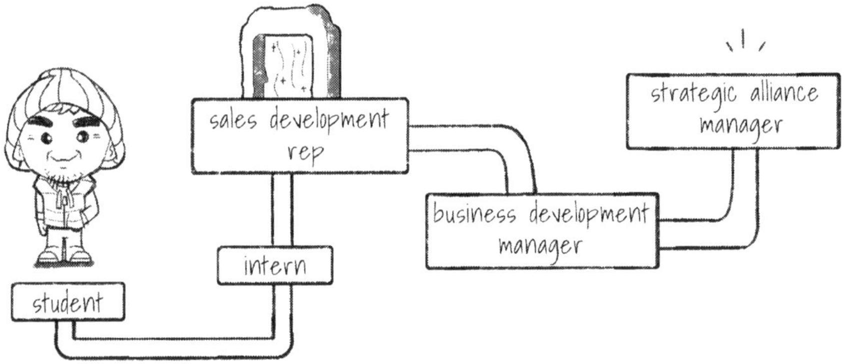

Alex, a Portland native, started his career as an intern at ESPN radio and afterward moved into radio sales before joining the team. Part of his SDR responsibilities involved running the marketing booths at software conferences, sorting through marketing leads that came in, and going to employee happy hour events.

Some of the leads that came in that Alex worked with were from other software and technology companies, except the titles were not from IT. For example, they had titles ranging from channel account manager, business development, and even CEO/owners of companies.

Part of Alex's responsibilities was to ensure the leads were routed properly to the right team. Through a combination of happy hours and meeting the teams responsible for particular partnerships, Alex learned more about other sides of the business.

Technology companies must have different ways to generate growth, and Alex had to learn this.

To help simplify what business development is here are the different ways to generate sales that Alex was learning:

1. Cold calling & emailing to reach people with zero relationship or awareness.
2. Generating leads through marketing to build awareness to create leads for those who may be interested and build a funnel to capture their information for follow up.
3. Working with other companies that you have synergy with and working together to share leads or help each other make introductions.

And it is that third way of developing business that Alex was working on. That third way is what business development also calls channel sales, or simply partners. In channel sales Alex supported and built relationships with software partners, as that team was dedicated to prioritizing the right partners for the company.

This role requires a different set of skills than sales, so Alex had to learn and develop additional skills. Because it's more about relationship building and involves a lot of happy hours and events with other partners within the community, that's what he had to gain mastery of. Alex also had to learn how to engage in strategic internal conversations to ensure that the right strategic partners were in place.

If your personality is one who enjoys meeting new people and love building relationships without the stress of carrying a quota the business development (aka channel, partnerships) may sound like a good career path to take!

Parker: Route to Marketing

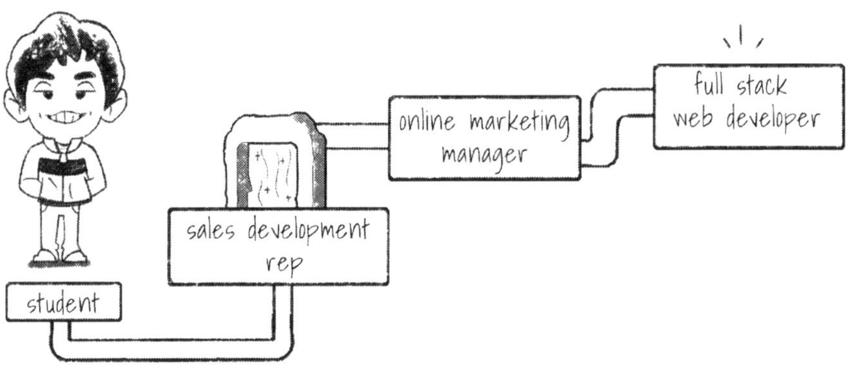

Parker, a North Carolina native, had zero sales experience and recently graduated with a history major. During Parker's tenure as an SDR, he found a great way to contact all of his marketing leads and prospects over a large region of accounts. He was able to craft emails that looked very targeted but were sent to hundreds of recipients at a time. This caught the attention of Andrew, the marketing manager, who was not happy with the approach. He felt that his email marketing solution in combination with Parker's emails would end up overlapping and upset customers, especially since he was not the one crafting the emails that Parker was sending out.

But Parker produced results! For example, Parker emailed a customer who was just about to sign with a competitor. That email landed in the decision-maker's inbox. Parker set up the meeting with the account manager, and they signed on as a customer that very month!

Eventually, through discussing his tactics and learning from Andrew, his marketing manager, Parker was eventually recruited to join the marketing team to help craft and manage the email automation side of their business. He would work with email workflows and drips, and would nurture leads within the system or send email blasts to prospects and invite them to local events at the nearby San Francisco Giants stadium. Just recently he informed me he learned coding and is now a full stack web developer!

I know that a lot of readers will be interested in breaking into marketing (note: the next chapter will cover all the different segments of marketing). I've seen a lot of marketing folks who started as SDRs in the B2B space. Again the idea here is that even if you don't see yourself as a lifelong salesperson, this gateway can allow you to break out to many different roles.

Next up, Karina!

Karina: Route to Account Management/Renewals

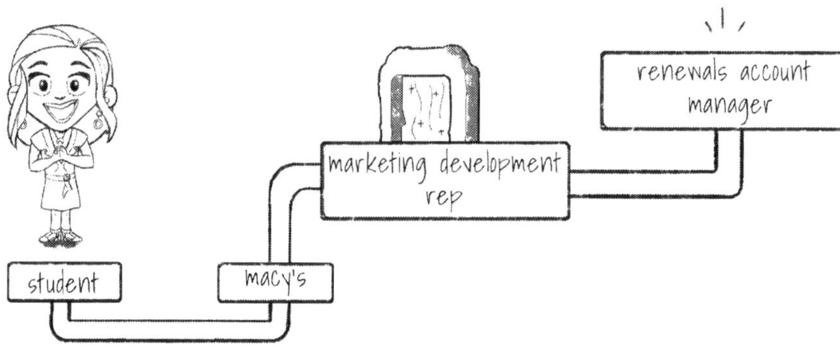

Karina, the youngest of the four and a University of Oregon graduate, worked in retail at Macy's, helping style customers. Karina was extremely nervous about moving from retail into tech. To help ease her anxiety, she was initially placed as inbound SDR (Also called an MDR: marketing development rep). What's the difference between inbound versus outbound? Instead of hunting for net new customer meetings (outbound), she was tasked with following up marketing leads that came inbound on a proactive basis.

After watching her teammates move into account executive roles, Karina knew that she didn't want to go that route. What Karina excelled in, despite being in a sales role, was proactively meeting up with folks in the front office, IT, finance, marketing, and newly hired teammates. She would invite them to lunch or happy hours, and start building friendships. As her friendships grew, she got to know a few folks who were building out the renewals side of the business. The renewals team's responsibility is to help make sure that existing customers are happy with the company's solutions and to ensure that they renew their accounts and potentially upgrade their usage of products.

If they are not happy it would be Karina's responsibility to help loop other people within the company, for example, support or the customer success team, to ensure the customer is successful. As she learned about this side of the business, she loved the concept of working with an existing customer and supporting them in their success as well as the thrill of driving a deal without the stress of carrying a new business quote that the AEs endure.

Karina knew she didn't want to be a sales rep but didn't quit. Instead, she was curious and eager to learn. She did her SDR responsibilities and without compromising her performance invested a fraction of her time into building relationships with other departments that led her to her main career today in renewals.

The SDR gateway is a great career launching point. You learn the technology and how to articulate it to customers. As long as you do the work, you'll have choices. Kevin continued his journey, moving on to be a successful enterprise sales rep. Parker found his niche in marketing by honing his email and writing skills. Alex found the channels route fit his personality, goals, and duties. Karina learned from many employees and found that renewals were the best fit for her.

Within the Sales Gateway: The Renewal Account Management (AM) Gateway

It is much easier to get a customer to spend more over a period of time versus getting new customers to buy from you for the first time! The amount of effort needed to create a first-time sale is the reason why separate SDR and AE teams exist. At a certain point in a company's growth, it is wise to build a separate team (the renewal account managers). This team ensures customers are happy and that you stay in touch throughout the year making sure they are aware of new offerings that exist as well as renewal of their existing software investments. This is great for folks who want to be customer facing without the pressure of delivering net new business! I asked my buddy Jeff Rachlin who built and runs the Cloud Security renewals team at Cisco. Here is a brief summary of some key things Jeff mentioned when looking at renewals.

Here are the two things you need to land a renewals account management position

1. You like talking to people.
2. You are good at solving problems.

You'll learn how to

- Manage teams and create your own operating process
- Save customers from canceling
- Upsell!

You'll get exposure to more sides of the business, such as

- Meeting with the product team and sharing customer feedback
- Working with the finance and marketing departments
- Working in tangent with the customer success team to ensure customers who are at risk

I couldn't believe how simple his approach was. I realized that renewals offer a very business-focused role. It has the aspects of sales where you are closing renewals. If you work for a company with a good solution, i.e., >80% renewal rate, you are focused on the troubled accounts that may need some tender love and care to help the customer get on board.

Sample Renewals Account Manager Career Description

Below I'm sharing a job description, so you can see how a technology company describes the gateway in their words. A lot of times first-time career seekers may not even have looked at these types of roles and responsibilities. I want you to see that often you already have the required skills to land the job. You just need to believe!

If you lack a requirement or skill, then see it as an opportunity to start learning and building that new skill right now. For example, if you are doing service-related work, like in retail or restaurants, focus on becoming a self-starter and salesperson; start changing and improving every day, so you'll grow as much as you can. In your free time take the actions described in section 3 to best position yourself for the many tech opportunities you'll be ready about in these gateway chapters.

Now for the sample renewals account manager career description. Check it out:

> This role offers the unique opportunity to touch all parts of the business. Not only does it require a high-touch customer interface, but it also puts you in direct contact with sales and support. You will also have the opportunity to leverage, showcase, and grow your account management skills, drive internal programmatic changes to better service customers, and be part of an amazing team of fun, energetic, and passionate account managers.
>
> This role is responsible for ensuring the customer knows and is aware of new offerings since their last year renewal, and the role is looking

to also upsell licenses, support, and/or new offerings. As the renewals account manager, you will ensure high customer satisfaction during the life of the customer's subscription, be responsible for managing ongoing post-sales outbound communication (non-support related), and partner with the sales and support teams to provide seamless customer interactions.

What You'll Do

- Execute all phases of the subscription renewal lifecycle to ensure the highest renewal rate possible
- Find upsell opportunities within our existing customer base to pass to our sales team in order to expand our relationships
- Generate invoices and work closely with finance to ensure payment is collected
- Be responsible for meeting renewal goals

Minimum Qualifications

- Excellent verbal (via telephone) and written (via email) communications skills
- Highly detailed oriented with the ability to manage a large volume of accounts, tasks, and responsibilities
- Strong customer service, listening, and presentation skills
- Bachelor's degree or equivalent experience
- Experience with SaaS-based applications and services preferred
- Experience working with Salesforce.com preferred
- Self-starter, high energy, positive attitude, customer-focused

Three Tech Novices' Spectacular Career Trajectories

To help shed light on how this role is a gateway role into other career paths here are short stories of three account managers who started together and the different career paths each grew into.

At one point in their careers Tess, Mona, and Miguel worked the demanding business of consumer retail (GAP and Anthropologie) and eventually found their way to account management.

Tess: Route to Training and Development

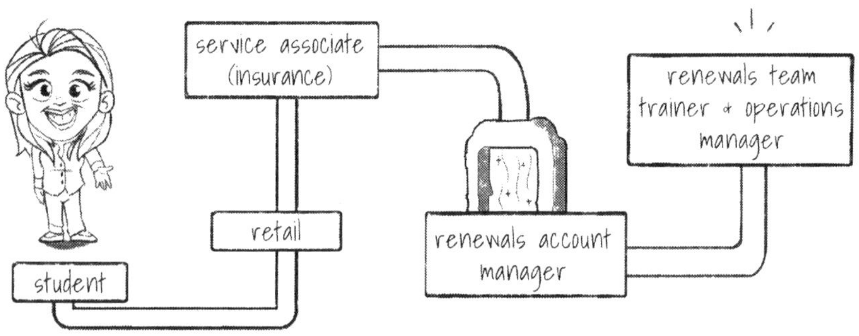

Tess heard about the account management role through one of the existing account managers. She was able to show that she could meet the two main criteria that the hiring manager, was looking for:

1. You like talking to people.
2. You are good at solving problems.

Tess couldn't believe that the interview process was that simple.

Key point: fast-growing SaaS companies that have a recurring business model will always need account managers to help renew and take care of existing accounts as the number of accounts will grow exponentially. Fast-growing companies typically have a higher risk tolerance in taking chances on candidates who they see can learn on the job.

As the company grew, Tess spent more and more time helping all the new hires to support the business. Even though the company didn't have a training program, after one year Tess and her manager worked on creating a role and process. That's how Tess created the training and development team.

Mona: Route to Sales Engineer

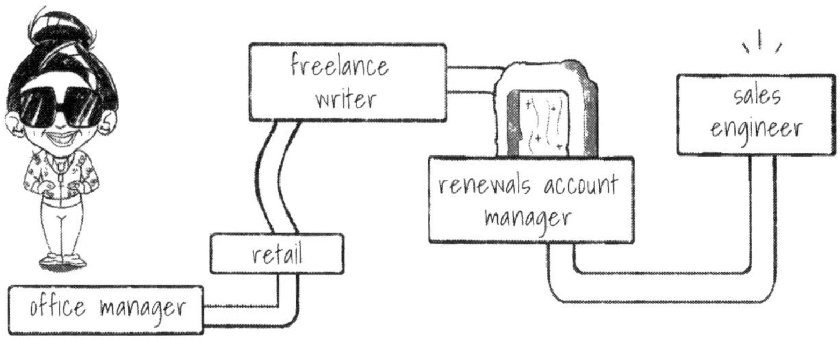

Mona was a writer at a Seattle newspaper company before her journey into tech. After sharing her desire to leave the cloudy weather of Seattle to move back to the Bay, her friend pointed her to a role at OpenDNS that she thought would be a good fit. Mona, like Tess, broke into the tech industry by starting in the renewals account management (AM) role. While in the role, she discovered that the pre and post sales experience were very different. In post sales she was dealing with customers who had already signed contracts, and the challenge of helping a customer before the sale (pre sales) intrigued her.

In technology sales every account manager and executive has a technical counterpart who helps with answering technical questions and troubleshooting technical issues that may come up during the evaluation phase. This helps take the technical portion off of the account executive's plate. These technical teammates are called sales engineers (SE). One of the coolest parts of being in technology sales is you have a tag teammate to help close a deal. It helps segment responsibilities, so the account team can focus on generating new business and closing it, and the SE team helps ensure the technical side is running smoothly.

Traditionally, most SEs have come from the support gateway (more details covered in chapter 10 The Customer Support Gateway). What's great about Mona is that when she was exposed to the presales side of the business, she started bugging her engineering friends at the company to help her get trained up on the technical knowledge. After learning a bunch of online resources and collateral, she took an online certification test and passed! When the opportunity arrived, she jumped into being an SE! This is another great example of breaking in, seeing what interests you have and where you want to be, and building the relationships with the folks who have achieved the results you are searching for and creating a plan to moving in that direction.

Miguel: Route to Customer Success

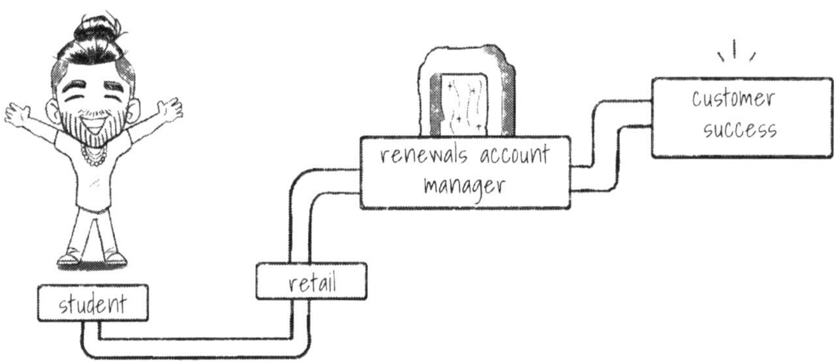

Like Tess and Mona, Miguel started on Jeff's team. In the renewals role he worked closely with the customer success team (more details covered in chapter 10 The Customer Support Gateway) on customer accounts that were having difficulties with the technology or were simply not adopting it. The customer success team is there to help prevent this. They either proactively reach out are brought in by the account managers to ensure that the customer is utilizing the product to the fullest, and any technical difficulties are sorted out quickly.

Tess, Mona, and Miguel's stories offer brief but powerful examples of how the account manager role works as a gateway for breaking further into the tech world and building a new career for yourself. By being on the AM side, Tess, Mona, and Miguel were positioned to be able to see additional paths into the tech world that they weren't aware of before.

The SDR and account management path allow you to break into the technology field even if you have little or no prior technology background. The SDR and AM path can also lead you to other career paths in tech like sales, sales management, supporting sales via sales engineering, marketing, renewals account management, and channels/partnerships. So while not everyone wants to be a career sales rep, sales and AM offer a fruitful gateway to other roles in tech.

In the next chapter, you will learn of a team that works closely with the sales team on a day-to- day basis—the marketing team! Marketing is the next gateway into tech that we'll explore.

GET YOUR FREE BONUS!

Thank you for reading & sharing this book!

To help with your **momentum!** Here are 4 **FREE videos**

I've included for you to bring the **book to LIFE!**

Go To: www.irwinki.com/p/FREETRAINING

If you want to see more videos!

Follow me: Search "Irwin Ki" on Youtube

Find me via @Irwin_Ki on Instagram

Get **connected with other readers** via FB group:

"Breaking Into Your Silicon Valley Community"

CHAPTER 4

THE MARKETING GATEWAY

What is your immediate reaction when you see the following?

And what is your reaction when you see this?

$124 billion dollars in market cap (at time of writing)! That's the difference in the power of marketing. It's why Apple can command lines of dedicated fans at their stores and charge the largest margin in the computer and phone business for their products, despite the fact that the cost of computing power has been going down.

Apple's marketing team crushes it! Marketing is probably one of the most sought-after fields. This can make it difficult to differentiate yourself if you have no prior marketing experience. Most people are able to transition into marketing via the SDR/MDR gateway (previous chapter). But that's not the only gateway in.

A great example of a success story in this area involves how one of Uber's executives joined the company. Here is a quick excerpt from the book *The Upstarts: How Uber, Airbnb, and the Killer Companies of the New Silicon Valley Are Changing the World*:

> "Austin Geidt's route into Uber was even more unlikely. She grew up in Marin, north of San Francisco, and attended the University of California at Berkeley ... When she finally ... graduated, she was adrift, insecure, and desperate for a job. After she applied for a barista position at the Peet's Coffee in Mill Valley and was turned down, she spotted a Tweet by Jason Calacanis about Uber. She followed a few links, then sent an unsolicited e-mail to Ryan Graves. He hired her as a marketing intern. "By her own account, she struggled to fit in. On her first afternoon the whole company went to Kalanick's apartment for a multi-hour jam session about the future of the company and the meaning of its brand. During the discussion, which lasted all evening, Geidt noticed that Kalanick, who paced the room, seemed to be really

> in charge. She found the experience incredibly stressful. "It was a bit overwhelming for me," she says. "I had terrible impostor syndrome and I hadn't even done anything yet. Over the next few months she was certain she was about to get fired. She had a poorly defined role that at one point required her to hand out flyers for UberCab in downtown San Francisco. Even the trivial act of writing an e-mail would send her scurrying to her older siblings for help and advice. At one point, Graves recalls, he counseled her in the stairwell to the office, with Geidt in tears. Instead of firing her, though, Graves gave her time to find her footing. Later he dismissed his first driver-operations manager and installed Geidt in the role. She would become one of the most important execs in Uber's early history."

WOW! It's safe to say that Austin's internship was a multimillion-dollar internship. This is a great example of how the marketing gateway can lead to other opportunities, even when starting so modestly. Austin's story also emphasizes in the point of the power of starting in a career in tech vs at Pete's coffee as well as the power of reaching out directly the decision maker (covered in chapter 16 & 17) Marketing is one of the most powerful forces for a business. You're not responsible for making the product, but you are responsible for making sure that product sells

Gateways

- Communications Coordinator ($49k-$78k)
- Marketing/Sales development representative (see previous chapter)
- Marketing analyst or intern ($15/hr to $83k)
- Field marketing coordinator ($47k–$63k)
- Public relations (PR) coordinator ($42k–$55k)
- Social media coordinator ($45k-$69k)
- Jr. Copywriter ($55k-118k)

Career Paths:

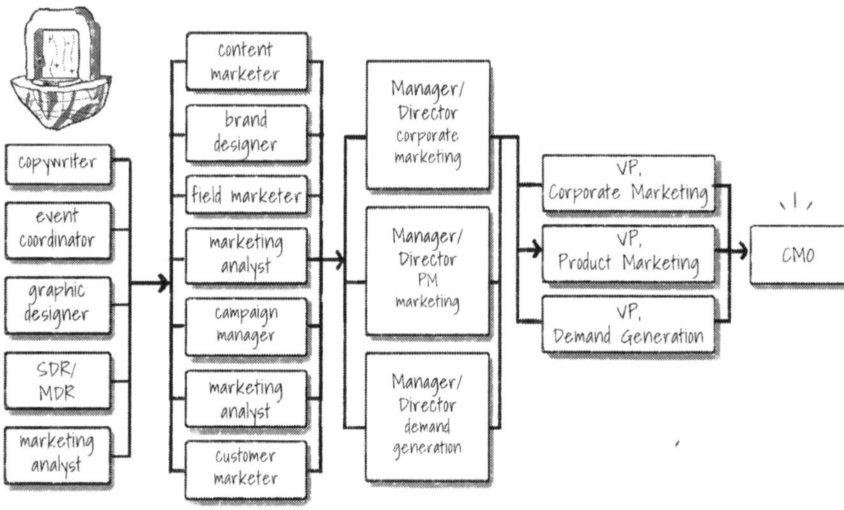

Marketing Simplified

- Corporate marketing: the business brand. It's what you see regarding Apple, their commercials, websites
- Influencer marketing: think Steve Jobs' keynotes when announcing the iPhone and new releases
- Product marketing: the words used to keep techies informed; think detailed technical info
- Demand generation: this entails meeting quotas, delivering measurable leads (email, online, conferences), and tracking how they convert

Marketing Skills Needed

- Corporate marketing: transferable marketing skills or promotions via below paths

» Influencer marketing: product and industry knowledge, plus presentation skills
» Product marketing: writer, transferable writing skills
» Demand generation: SDR, analyst, event management skills

Sample Events and Field Marketing Intern Description

In the marketing world, their gateway jobs can take the form of an event and field manager, marketing internship, marketing data analyst, growth marketing analyst, or digital marketing analyst. Below is a post that will shed some light on the responsibilities in the field and events side of tech marketing. You may want to pay attention to how "non-tech" this role is. A lot of SDRs from sales transfer into event and field management as lots of marketing leads are generated at trade shows. One thing you'll notice that is unique to this role is that instead of giving job requirements, they give "characteristics" and soft skills needed, which I've put in **bold**:

> As the events and enterprise Marketing intern, you'll help drive the events strategy by supporting our participation in trade shows, on-site client meetings and experiences, our annual client conference, webinars, and more. In addition to touching many aspects of events (working a trade show floor, crafting and leading the production of swag), you'll develop a webinar playbook to engage buyers and users at every customer account, educate them, and move them through the buyer's journey. Think you're up for the challenge?

Your Responsibilities

» Research events trends, upcoming tradeshows, and meaningful webinars put on by our competitors to make sure our events strategy is robust and state of the art

STOPPING ADVERTISING TO SAVE MONEY IS LIKE STOPPING YOUR WATCH TO SAVE TIME.

— HENRY FORD

- Collaborate with Communications to build a program for pre- and post-event buzz, including press releases, blog posts, emails, social posts, etc.
- Build out the quarterly skills webinar pipeline with content, speakers, and fresh ideas
- Organize event supplies and collateral and handle shipping and receiving for event-specific items
- Research industry topics, draft blog content, and proofread webinar communications and sales collateral
- Partner with Sales to arrange the details of onsite customer visits

What it takes to catch our eye

- **Organized** and **meticulous** with **excellent communications skills**
- **Passionate** about event planning
- **Self-motivated** and willing to go above and beyond to **meet deadlines with little supervision**

How to really knock our socks off

- Able to embrace ambiguity, anticipate project needs, and use an "out of the box" approach to **solving problems**
- Willing to work **occasional evenings on the night of events**
- Passed "Events 101" and developed beginner **planning and project management skills**
- Can **adapt to change** and jump in where needed
- **Think creatively** about how to stand out at events beyond swag and expensive giveaways
- Willing **to do whatever it takes to pull off a** successful event... events are not as glamorous as they might seem

Courtland's Take on Tech Marketing

To help navigate the complex world of technology marketing, I've had a chance to chat with Courtland Smith, senior director of marketing at GitLab, described by Business Insider as one of the "top 51 enterprise startups to bet your career on" in 2018. Courtland has been part of four startups, two of which were acquired. He's hired over seventy employees, and he's going to go into all things tech marketing.

Irwin: Can you talk a little bit about yourself as far as your background, how you got into technology, and how you got into the marketing side of business?

Courtland: Like many B2B [business to business] technology marketers, my path to marketing in the tech world has been unconventional. You don't go to school and learn about B2B tech marketing. If you learn about marketing at all in school, it's going to be consumer marketing. And it's going to be a marketing skill set that's better suited for large consumer brands than for a small technology startup trying to introduce a new idea or a solution into the market.

My background and how I got into marketing started with a lot of math and a lot of spreadsheets, working as a data analyst and a telemarketer at a company that does fundraising for charities. I was literally hooked up to an auto-dialer, making 120 calls a day and trying to raise money for charities. The opportunity just presented itself to take on some additional work, which was crunching through all the data, crunching the day's call logs to then put into reports for my company's clients. From there, I learned how to get good at Excel, how to do some light programming, and how to automate and make my job easier and more efficient. By necessity, I spent a lot of time self-educating, learning how to be better with spreadsheets, how to automate repetitive tasks, and how to turn data into insights for clients.

The skill that I happened to have was around data analysis in terms of putting together reporting for clients. So by virtue of that background, I was able to land a financial analyst position at a small Bay Area-based startup called Adify. It was very interesting. I reported to Rick Thompson who is a partner at Signia Ventures, very senior, and who has done a lot with his career. Rick was a really good guy to learn from, and I got some very interesting exposure to the startup world at a pretty early age and pretty early on in my career.

Because it was a startup, 50 employees at the time, everyone wore a lot of hats. I found that, unlike many larger companies, at a startup, if you can do something, if you have the skill, if you can add value for customers and add value for growth and for sales, then there's nothing stopping you from doing that, even if it's not technically in your role description.

I applied my skills with crunching through large sets of data, in this case, ad-server data for Adify, which was an ad-tech company. I did this, so I could present information to clients in terms of how the networks and websites that they were managing were performing and how the online advertising campaigns running against the sites within those networks were doing. This helped them determine what publishers and their networks would need to be paid, things like that. I got to work very closely with product management in that endeavor, and I got to figure out how to go from working with information in a spreadsheet that was emailed to a client, to actually turning that information into a product and incorporating it into the logged-in experience so anyone at any time could see all of the metrics in order to evaluate the performance of advertising campaigns running on their networks.

In addition to that, I worked very closely to the marketing department at this company, and I quickly gained the reputation as someone that knew how to use Excel very well and that could take big sets of data and present them in interesting and insightful ways. The marketing department at Adify tapped me to collaborate on a project to go through all ad data from 170 different networks, tens of thousands of publishers across many different verticals, to explore trends in the online advertising industry, so that the marketing department could publish a report and deliver that report for PR and thought leadership purposes.

The CMO at the time took my findings and figured out how to plug them into a broader interesting story about how online advertising rates in real estate-focused websites actually worked as early leading indicators that there was going to be a recovery in the real estate market.

Irwin: Really?

Courtland: Yes. I remember at the time for Joelle Kaufman, my boss and CMO at Adify, it was like music to her ears when she saw that the *New York Times* had picked up our report and picked up the data that I had mined. It got published in the *New York Times* to a national audience. This translated as massive attention for our little startup from the *New York Times*, their online property as well as the newspaper. Our startup was able to be tied into that bigger, more interesting story. From there, Joelle offered to bring me on to the marketing department.

One of the things I always struggled with as a new grad with not a lot of experience was I always felt like I was a jack of all trades and a master of none. I was pretty good at writing. I liked to analyze data. I liked to automate things. I thought, "What is my unique strength? What am I uniquely best at? I'm not the best automator, I'm not the best finance person, I'm not the best project manager, I'm not the best writer. But I can actually do all those things pretty well." I came to the conclusion, using a career advice tool called Calling Cards, that my unique strength wasn't that I was good at any one thing; my unique strength was that I was pretty versatile and okay at many different things. For that reason, I then went on to explore areas where I could apply my jack-of-all-trades background, and that was going to be my key strength. I sought out fields that really value people who can do a lot of very different things as opposed to hyper-specialization. Marketing came to mind.

Next I spoke to the people at Adify. In so doing, I found out that marketing wasn't just one discipline; marketing was 15 different disciplines all encompassed under one umbrella. Each discipline is different and requires different skill sets. Multiple disciplines are a sub-discipline of marketing and requires expertise in terms of left-brain and right-brain thinking, creativity as well as the ability to be analytical and data-driven. For that reason, I was very receptive to joining the marketing department at Adify.

Irwin: What are different ways that you advise people to break into tech companies? What are the different gateways in which people can break in if they are interested in marketing, particularly tech marketing?

Courtland: As I already noted, there are a lot of different disciplines within marketing that require different skill sets. With B2B tech startups, their marketing departments are not looking for an experienced candidate. They want someone that has a skill that they understand well and can apply to achieve higher and faster growth for the startup. That's the name of the game with startups.

Sales development representative, or SDR, is the first gateway I want to mention. The skill of understanding customers, knowing their language, understanding the market, understanding how to talk about your product in a way that compels action, that is absolutely critical for a marketing department. As a sales development representative, you can expect to be sending emails and managing the flow of those emails in a little bit of an automated way using Outreach or Yesware. The bottom line is that with automation you can continue communicating with a higher volume of people in a relevant way while at the same time managing a larger volume of leads for follow-up and prospecting.

Many people starting out in SDR make the mistake of not being able to scale what they have learned. Once you figure out how to get a response from this or that company, you need to ask yourself, "What if I could apply what I've learned at scale? What if I could use the insights that I have learned and then apply that, instead of one to one, one to many?" This could mean running programs that have a larger impact on more people and a larger audience. It's a great background.

I certainly recommend looking into a sales development role if you want a role in B2B marketing because you learn how to talk about the product, you learn about the market, and you learn about your customers. That is of absolutely critical importance and value to any B2B marketing organization. But there are other paths; that's certainly not the only path and not for everyone.

Working at a PR agency is another great gateway. I've hired around 80 people in the past six years, and one of the things that I look for in content marketers is a background in journalism, working at a PR firm, or working at a publisher. I've looked for people that have demonstrated strong ability to execute under pressure and to write at a very high quality, clearly, and concisely. That is a valuable skill set, just like knowing a target market or a customer. That's a skill set that can be immediately applied in a B2B marketing organization, specifically a content marketing organization.

That would be the second path that I'd recommend.

The third one is more of the path that I took, which is finding the analytical role, working in the finance department and getting comfortable with large sets of data, and figuring out how to draw insights from it that can inform strategy and decision-making. Every marketing organization is very concerned with strategy. They want to know if the messages that they're taking out to market are working. Are they resonating? Are people responding? Whether it's an online marketing role or you're running online ad programs or paid search programs, or an SEO marketing role where you're trying to figure out what keywords are important to your audience, these questions are critical: how competitive are they? How much traffic do they generate every single month? There is a lot of appetite within marketing departments to find people that have a proven track record of being able to effectively analyze data, turn that into insights and action, and inform marketing strategy.

Irwin: If you're a great writer you can find a way to apply that skill in tech marketing. Definitely content marketing is always needed. Also, as you pointed out, there's the option of going into tech marketing through an analyst-type role. Any other gateways that you can think of?

Courtland: The last one that would come to mind is event management. There are many industries out there that need to manage events and create compelling experiences that are live for people. Just as the hospitality industry has event managers, many different industries rely on event managers. It's a skill set can easily be parlayed into a role in field marketing. A big component of the field marketing person's responsibility is going to be local and regional marketing events, so that means creating those events, managing them, getting people to come to those events, and making sure that when they do, they have a great time, have good conversations, and learn about your company or product.

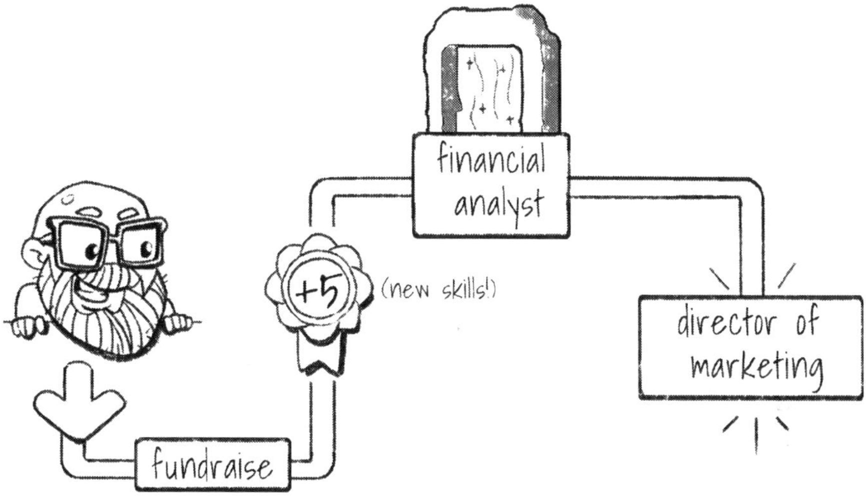

Irwin: That's a great gateway. For events like Dreamforce, big enterprise companies have a huge conference. And not only that, there are security conferences all over the world happening every month, so having someone to help organize that is a great way into the tech community. I think that's a great idea.

Courtland, you are a master of this field 100%. I'm learning so much as we speak. Are there any other questions I haven't asked you that you think I should ask?

Courtland: I'd say, what to look for when thinking about the startup that you want to join. How do you evaluate a startup?

Irwin: How do you choose the right startup for you first career move?

Courtland: To begin, I think it's better to get started somewhere than nowhere. It's okay to not be that selective for the first startup that you join because you will learn and you will get experience. You will be able to take that experience into your next job if you do well and you learn a lot.

Think of your first startup as a paid education that you can then use to get a job in a company that you're more interested in. Most success in startups has less to do with the company and more to do with the market that that company happens to be in. Ask yourself, "Is that market doing well?"

You also need to think about what is exciting to you. Network security, I think that's exciting. I think that's cool. Not everybody thinks network security is cool.

Startups can be demanding. You can work long hours without a lot of guide rails that you have at larger companies. You might find yourself having a tough time balancing your life and work. You're going to be sprinting especially early on in your career when everyone is really eager to stand out, learn as much as they can, and work as hard as they can to advance their careers as quickly as they can. It makes it so much easier if you're passionate about what your company is doing as well as its vision, mission, product, and/or problem that it's solving. I think that's certainly key.

The economics behind a company matter too. It's import to understand what their renewal rate looks like. And you should ask these questions in interviews.

Irwin: What are the questions?

Courtland: What does your renewal rate look like? What's your average deal size? How long does it take to close a deal on average? When you ask these kinds of questions, you'll find out whether the company knows what it's doing. Do they have a good pulse on their business? A good marketer should be able to answer these questions.

Again, the world is full of mediocre marketers. You have to have a high bar when thinking about who you want to work for and who you want to learn from. I think it's important to understand and read customer testimonials to determine the value customers end up getting from using that company's product. If they don't have those, it's a big red flag. Maybe it's not a place to pick for one of your first few companies.

I think it's also good to understand how experienced the executive team is. There's a lot of great founder CEOs out there; I've worked for a number of them. But it's good to know that there's a high bar for the executive team. Executive positions matter. They have a lot of influence on the success or failure of a company as well as the relationship that the board of directors has with the executive team. You want a healthy company where people that are in their roles are qualified. There's a lot of companies out there that suffer from cronyism; somebody might be in a really senior role just because they're friends with the CEO. Those kind of companies you're going to want to avoid like the plague. That's not how a healthy, sustainable company culture gets built. That's how a toxic political culture gets built.

I'd certainly look at reviews on Glassdoor.com, the online community where candidates can anonymously post their interview experiences, compensation they are paid for specific roles at a company, as well as general reviews of what their experience was like working at a company.

Most of the salary numbers posted in this book are from the salary info on Glassdoor.com. Also, know that bigger companies that are well capitalized will pay exiting employees to keep quiet. Back to the matter, they'll have you sign a non-disparagement clause to protect their reputation.

Interview Gold Nuggets

Courtland came to me highly recommended from the CMO of Cisco Security, and he certainly lived up to that high rec! He offered so much wisdom in that interview. I've compiled this list of the crucial takeaways:

- Marketing that is taught in college is mostly for the consumer market. A lot of marketing in the B2B space breaks from that in unconventional ways.
- There are multiple gateways into tech marketing:
  - SDR
  - Marketing analyst/intern or adding value to the marketing department; in Courtland's case, being good at data
  - Field marketing—running trade show booths
  - PR experience
  - Event management
- Your first startup will be a learning experience; just get in the game!
  - If you want to increase your chances of finding a successful startup—look at the leadership team, ask questions around their numbers to see if they are growing, and ensure the startup has a large, addressable market.

Marketing is a high-needs area to start your career in. Plus, there are so many online courses on how to build up your social media and paid advertising skills. Getting knowledge through the online courses will add value and help you as an applicant to stand out. Remember, it's the journey, not the destination. It's who you become in the process.

A Quick Rant on Social Media Trends

A lot of these online marketing courses have also resulted in folks skipping working for a company to building out their own digital marketing business.

Why?

More and more attention is going into the social media space. Think about it. How much of your time is spent on Instagram, Facebook, or LinkedIn versus traditional media like TV, magazines, or newspapers? More and more attention is spent on the former platforms, and most business owners have NO IDEA how to take advantage of the new shift.

This knowledge creates a gap. If you are interested in marketing, then the first thing to do is get in the game! Either A) work on breaking in and building a career and learn while you are getting paid and/or B) learn from people who have the results that you are looking for, like I did with Courtland. There are so many masterclasses on starting your own social media agency with the success stories to go with it.

It used to be that coding was the most valuable skill a decade ago. In the next 10 years digital marketing going to be the most valuable skill.

Now, let's dive into a very important tech gateway, the one that ensures a business runs smoothly and efficiently—the operations and analyst gateway. Plus, the operations and analyst gateway is great for folks whose personality may not fit the sales or marketing personalities.

CHAPTER 5

THE OPERATIONS AND ANALYST GATEWAY

Not everyone is made to be in the spotlight like sales or has courage to yell on a megaphone like marketing. Greatness can't be achieve without a process and ensuring that things are running smoothly!

The next gateway of interest is the operations and analyst gateway. With this gateway the smaller the company, the more roles and responsibilities the operations person will carry. The larger the company, the more defined and honed the role becomes. Why? Because the more a business grows, the more segmented and complex it becomes. As a result, the operations people's particular roles will reflect that.

Here's the best part regarding these gateway roles: there are many operations-type roles opening up all the time because as businesses grow, their needs change. The changes require bringing in more systems; these include more software, systems, support HR, IT, sales, etc. When you bring in systems of this nature, they don't integrate immediately. There will be areas where executives will need to pull data and reports to ensure everything is running smoothly. And that's where the operations department and analysts come into play.

If you are detail-oriented and love to figure out how everything works and comes together in a business, this area is for you. With this gateway, there is a lot of job security, but be aware that it will require creativity to stand out if want to move on or get promoted.

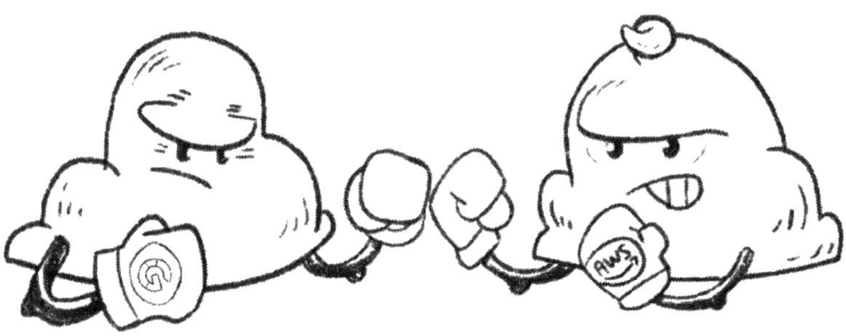

Gateways

- » Sales (operations) analyst, contractor, or intern: ($49k–$82k)
- » Product (operations) analyst, contractor, or intern: ($54k–$90k)
- » Systems analyst, contractor, or intern: ($53k–$93k)
- » Finance or data analyst, contractor, or intern: ($49k–$87k)
- » Marketing analyst, contractor, or intern: ($45k–$88k)
- » IT analyst, contractor, or intern: ($46k–$85k)
- » Facilities coordinator (44k-107k)

Career Paths:

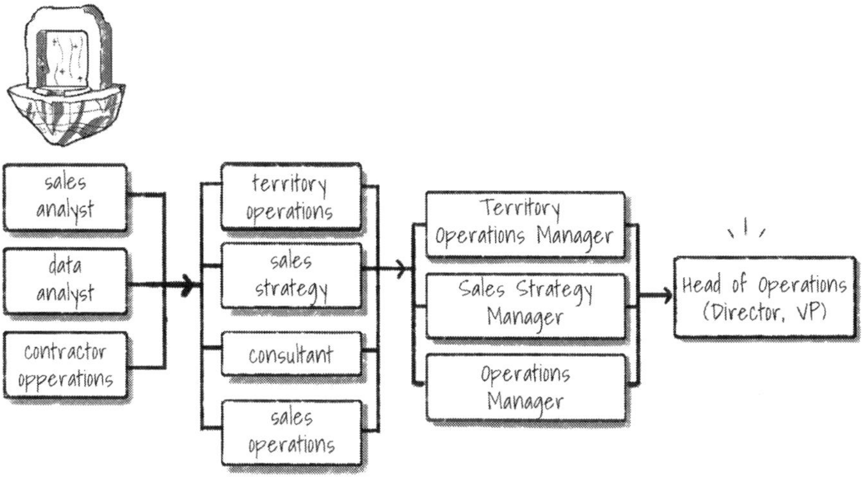

Operations Simplified

» Sales analyst: help the sales team automate their backend sales process

» Product operations analyst: responsible for day-to-day blocking and tackling to administering how the product and customer interact

» Systems analyst: help with internal software for your team; ranges from company to company

» Finance or data analyst: have to be good at Excel, Numbers, or finance, as well as looking at patterns to identify possible efficiencies (see chapter 6: Finance Gateway)

» Marketing analyst: (see the previous chapter on marketing)

» IT analyst: help the IT team automate some of their manual processes and create ways to automate into other existing and new systems.

Skills Needed

- High attention to detail and superb organizational skills
- Great appreciation of processes and rules
- Particular analyst tracks require specialized knowledge in the product or systems to help you stand out:
  - Sales analyst: customer relationship management (CRM)
  - Finance analyst: Excel, SQL
  - Data analyst: data software, i.e., Tableau, Excel

Sample Sales Operation Analyst Career Description

Let's take look at what an Operations Analyst will do and the type of person you'll need to break in. Don't be intimidated by all these responsibilities. If you are detail oriented and is ok not being in the spotlight. I think the interview after this description will help give you hope in breaking in via the operations & analyst gateway. As usual I've **bolded** a few things I think you'll find interesting.

What you'll do

- Develop a comprehensive **understanding of product strategies and financial initiatives to align our sales leadership** teams through **data driven metrics.**
- Assist in the development and **reporting of key sales performance metrics** to help drive increased sales productivity. Distill reporting into presentation format for sales leadership roles.
- Assist in the development of annual and quarterly forecasts; create sales leadership facing forecasts, pipeline analysis and trending.
- Partner with the global sales team to help **drive education on sales processes,** operational updates, and best practices to help build increased seller productivity and adoption of internal systems.

- Other ad hoc responsibilities in support of sales operations and sales finance.
- **Salesforce.com Administration:**
  - Day to day management of Salesforce configuration – customizing core architecture as well as profiles and user permissions
  - Manage and respond to end-user support and customization requests
  - Manage and customize dashboards and reports for various teams

What you'll need to be successful

- **Intellectual curiosity** and a "figure it out" mentality
- Impeccable **organization skills** and attention to detail
- Expert in Excel and PowerPoint
- **Salesforce.com Certified Administrator** with 1-2 years of hands on experience
- Experience in sales finance or sales operations

What I'm constantly amazed by is the diversity of personalities and people in the world. I'm super thankful for teams that help build the backend processes and automation so that I can focus on my responsibilities and goals. These roles are always growing and in demand. As a company grows, scales, and acquires there are more and more systems and processes to keep the organization in check. Ross has an amazing example of this. Check out his journey!

IN THE END, ALL BUSINESS OPERATIONS CAN BE REDUCED TO THREE WORDS: PEOPLE, PRODUCTS, AND PROFITS.

— LEE IACOCCA

Ross's Take on Sales Operations

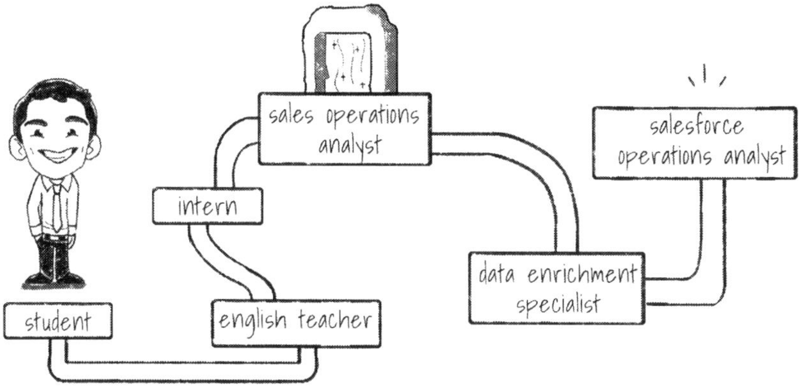

As fate would have it, I ran into my Ross Nichols while I was touring the Salesforce.com office. Ross is currently a sales strategy and operations analyst at Salesforce.com. He used to be a school teacher and moved to San Francisco, on a whim, with *no job* locked down or prior tech experience. I also love his story and how he explains that operations is a lot like maintaining a car. Let's dive in!

Irwin: What's your story, Ross? What did you study in college? How did you decide, "Hey, I want to break into technology"?

Ross: My journey to tech has been pretty roundabout. I studied economics in college, where I basically just read a lot of books about economic history and how theories developed. I didn't really know what I wanted to do with my life after I graduated. So, I went and taught English in South Korea for a year.

I really enjoyed being able to travel, and teaching was actually a lot of fun too. Getting to work with little kids. But one day a friend from college who grew up in San Francisco, in the East Bay, sent me an email, out of the blue, asking if I wanted to move to San Francisco and take a spot in an apartment that he was looking at. And so I replied, "Sure."

Irwin: And you had no job lined up?

Ross: Without any job.

Irwin: And why did you move to San Francisco? What called you?

Ross: It was more just really good timing. I didn't really have anything lined up, but I knew I didn't want to live in South Korea. I wanted to get back to America. I started looking for jobs and quickly realized that I was frequently asked, "Can you code?" and "Do you want to sell?" Basically, those were the two channels that were obvious from the get go. Because I didn't really want to live on commission, and I didn't have any coding experience, I remember thinking, "It's going to be a little bit harder to find a job than I thought."

So, I actually started out at a temp agency to see what was out there. And that's actually how I first ended up at OpenDNS. They were looking for someone to do data entry, manual-type stuff.

Starting out, I was a temp worker (also known as a contractor) for two-week time slots. My contract only lasted for two weeks at a time. If the company needed more work at the end of the two weeks, I kept working. Eventually, the company was looking for someone full-time, and they moved me to a full-time contract instead of one that expired after two weeks. That's how I got moved from a temp to a contractor role. I still didn't have employee benefits because I was a contractor, not an employee. Eventually, my full-time contract position led into a full-time employee position with benefits.

I started out with the manual data quality type stuff, working mostly out of Excel spreadsheets, making sure that the stuff going into Salesforce was as clean as it could be. And then it got more into working in Salesforce itself and progressing from getting the data ready to enter into Salesforce to actually managing the data once it was in Salesforce.

Irwin: So, Salesforce operations?

Ross: Yes.

Irwin: What do you think are the qualities that you were able to show to become their ideal full-time candidate?

Ross: I think one was just being able to use Excel, not even at an advanced level, necessarily. I knew how Excel worked in a very rudimentary framework.

Being able to pay attention to detail was also really important. And being able to meet deadlines because they were always aiming to get stuff into Salesforce as quickly as they could. And so it was, "Here's a list of leads or contacts that we need verified. Can you go through them, make sure everything looks okay? Get those uploaded?"

Irwin: What do you do now at Salesforce? Can you give the story about going from OpenDNS to Salesforce?

Ross: I've been with Salesforce for going on one year. Now I work on the sales strategy team, which is sort of an umbrella term that encompasses a lot of different functions. We have people who specialize in dashboards and metrics, presentations to our CEO, CFO, and other cross-functional teams that we work with. Also, we have the operations side of making sure that our AEs are able to continue to do their job as smoothly as possible.

I actually work for Salesforce.org, which works specifically with nonprofits. Because we work with nonprofits, we have some eligibility rules that stipulate we can't sell to everyone.

Irwin: What do you like about working at Salesforce?

Ross: I actually was asked this question when I was on a panel speaking before middle schoolers. They wanted to know the same thing, and it's basically the people. The people are awesome. Everyone is always upbeat, willing to be collaborative. We're very transparent here. There's not a lot of hidden agendas or politicking. It's nice to be able to come in, do your job, and work with people who are passionate about what they do. So, that's why I love Salesforce.

Irwin: What do you love about technology?

Ross: When I moved to San Francisco, I didn't really know anything about Salesforce. Now I basically live in Salesforce. All day, every day, working in Salesforce. I've come to appreciate how powerful of a tool it is. And being able to work with nonprofits in schools and organizations that are trying to really do good in the world and helping them leverage the tools that we have—it's a really good feeling.

Ross shared some great insights with us. He mentioned that you can start a career in sales and then move to an operations role on top of starting directly in operations. He also mentioned two of the skills that you need if you're going straight into a sales operations analyst role:

(1) Excel skills and (2) a Salesforce administrator certification.

A lot of companies today have a Salesforce CRM. An effective way to stand out is to invest in your own education into a Salesforce role.

Irwin: What advice would you give someone starting out of college who wants to go into tech but may not want to start in sales? You did come in through a temp role, which is great. But that requires a little bit of luck. If you had to do it all over again, what are the roles that you would look for? And what are the entry level gateways into a tech organization?

Ross: My case was pretty atypical. I had to get very lucky. But we have people on our team who are Salesforce certified. They're very technical, and they know how Salesforce works.

If you start out as a BDR, SDR, salesperson, or account executive and decide you don't want to do it anymore, there's always the option of going into a sales support role, not in the sense that you're qualifying leads or anything, but developing sales strategy to really enable salespeople to do their jobs. And so, having the perspective of a salesperson, you're really in tune with the needs of what a salesperson needs.

There are also people with MBAs on our team. They have gone to business school, and they understand how business works. But they haven't necessarily been a salesperson. They might not necessarily be the Salesforce expert, but they really know how to talk to the managers and executives we deal with on a regular basis. The nice thing about operations is that there are a lot of avenues within it.

Something else I want to add . . . *Zen and the Art of Motorcycle Maintenance* is a book that has stuck with me. The author is really into motorcycle repair and motorcycle maintenance. It got me thinking about how a sales organization is like a car. You start out with a skeleton of a car, nothing except the frame. In a business, we have different teams that, in theory, make up parts of the "car." So, you have the Salesforce administrators and developers who "build the engine." Without them, you can't go anywhere. And then with sales strategy, we're sort of like the "mechanics." I envision myself in this scenario like we're always listening to the engine and making sure it is running properly. We are asking ourselves if there is anything that might be off or if there is anything that we can fix. In sales strategy, our job is to make sure we're performing as efficiently and effectively as possible. If the engine is broken or it's not working to its potential, we tinker with it. We figure out what's wrong and make sure it gets back to humming along.

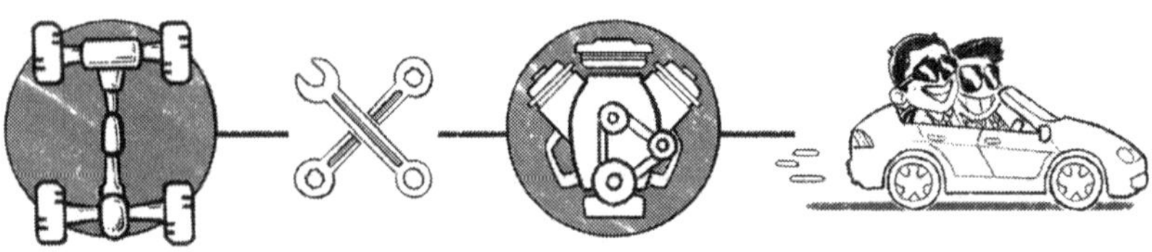

You have the salesperson who's driving the car. They're driving because they're the ones that are interacting with the customers; they're the ones who are generating revenue. It's awesome if you have a really nice car that has a perfect engine, but if you don't have anyone to steer it along, so to speak, then you can't go anywhere.

As far as the ideal strategy candidate, it's got to be someone who views themselves as a mechanic, someone who really wants to understand what goes on behind the scenes, who has a technical mindset, but is able to communicate a potentially complex issue and explain it to an AE, a manager, or a VP who doesn't necessarily have the background knowledge. They just want to know in as few words, as simply as possible, what to do to improve or what they need to be doing. As a strategy person, you need to be able to understand what's going on. If someone asks you a question, you need to be able to understand how to answer the question they're asking, maybe clarify what they actually mean if they don't phrase it clearly.

Irwin: Actually, I never thought of it like that. I love the simplicity of that example. In a car, where sales are the car, you have the operations and analysts, which are the engine because they're essentially keeping the CRM, which is what Salesforce sells, the automated software to track everything.

But a sales strategy is the mechanic that actually listens and determines if the engine is running well. Then, they report back or make tweaks to the car, so it runs faster. They communicate to HQ, whoever the driver is reporting to, and say, "If we make these tweaks, invest money here, or pull some levers, we can get more out of a company or more out of the sales reps, faster speeds." A mechanic kind of mindset. I love it. That's good feedback.

If you were 20 or right out of college, any advice other than Excel and Salesforce experience?

Ross: It's easy to get overwhelmed. I definitely felt that I was in over my head at times, but it's okay if you don't know what you want to do right out of college. Unless you're studying law or studying to become a doctor, there are lots of different things you can do. San Francisco is awesome because we have such a wide range of companies to choose from. There's the startups like OpenDNS used to be; there are the behemoths like Cisco and Salesforce; and there are companies everywhere in between. When you're thinking about your future, you can think about the type of job you want to do, but you can also look at your options for working for a startup or a large company.

Irwin: What have you noticed are the big pros and cons of working for a startup versus a huge company, a publicly traded company?

Ross: Pros for startups: flexibility, the potential to work for the next global behemoth if you get in on the ground floor of a startup that ends up being the next Apple.

Irwin: You're a millionaire.

Ross: Yes. However, the trade-off is you usually make less money for the potential to make a lot more money later on with startups. If you want to test the water, it is really exciting. There's a lot more uncertainty about it, which is probably the biggest con in that you're always fighting for funding when you're a startup.

Irwin: What about for a big company?

Ross: For the big companies, you don't have a lot of that risk because you are a public. If you are working for a big publicly traded company, you're accountable to your shareholders. You can't just do things on a whim that you might have been able to do as a startup. There are processes that seem trivial, menial, or just don't seem to make a whole lot of sense, and a lot of the times it's because there has to be a system of checks and balances because, regardless of whether you're the CEO or you're at the bottom of the totem pole and straight out of college in SDR, you're contributing to the company's bottom line and the company's performance. Ultimately, you're accountable to the people that own the company.

* * *

The operations and analyst roles are much needed within a growing organization. Ross was a teacher and moved across the country with no job in San Francisco and was able to break into tech! It's because there is a huge need, and he was the right person for the role. A lot of companies are using more temps and contractors to fill the manual part of their business. Take Uber and Lyft, for example, their entire driver force is composed of contractors!

A lot of times contract roles get your foot in the door. The operations roles are great to help ensure that processes are being executed, and analyst roles are great to see how different processes can be optimized. If this sounds like your cup of tea, this chapter points you in the right direction.

At the end of some of the processes, there is going to be income and expenses. The next chapter offers a lot of great content from folks who explored the traditional banking side of finance and found there home in tech. Technology has become more and more of a talent drain to the banking scene. Let's find out why!

CHAPTER 6

THE FINANCE GATEWAY

Let's talk about MONEY!

Most folks who are interested in finance study economics, accounting, and business, and head off to pursue consulting, ibanking, and other opportunities at Wall Street. I've seen a few of my friends pursue this path, and they were left with long hours, a lack of fulfillment, and wondering if there is more.

The good news is there are a lot of transferable skills that are needed in corporate finance. I've seen many transition their careers from working from the hedge funds and banks to technology companies. You'll see in the stories in this chapter that finance roles are all critical for growth. The more a company grows, there is a team needed to help manage the receiving, spending, and optimizing all the different business units. Welcome to FINANCE!

What if you didn't go to school for finance? The good news is there is an entry role, the account clerk, also known as a bookkeeper. Day-to-day activities involve entry accounting, payroll, accounts payable, accounts receivable; in summary, you assist in keeping the financial books up-to-date. There are tasks that require bi-weekly executions, such as paying employees on the 15th and 30th of every month, ensuring vendors are paid, and collecting checks from goods and services sold.

For those of you already working in banking or consulting today, this gateway is too entry level for you. No problem! The next chapter covers the gateway that bankers and consultants typically use to break into tech.

Gateways

- Accounting clerk ($25k–$49k)
- Finance analyst/internship ($60k–$102k)
- Finance/MBA graduate

Career Paths:

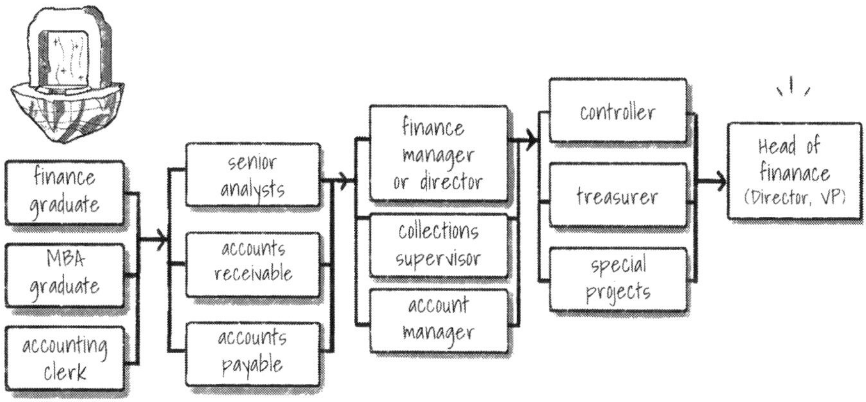

Financial Skills Needed

- Most of corporate finance will need a financial bachelor's degree.
- Bookkeeping experience helps for breaking into the entry-level roles.
- Excel data is typically needed. Analysis of the Excel data will set you apart.

There will be a finance team at every company, period. Money is a flow of resources, and there needs to be a team to ensure all resources are coming in and out in a timely fashion and that resources are utilized properly. If this is done incorrectly or if leaks in the business are never stopped, businesses can flop overnight.

If you are good with numbers, detail-oriented, and can problem solve your way to ensure the numbers add up, then this is the place for you.

With leadership skills added to the mix, you could become a chief financial officer (CFO). I've noticed that the CFO is one of, if not the best, places to be in technology.

Why?

Most entrepreneurs and their investors want to find a financial exit through acquisition or an initial public offering (IPO). What better way to have access to the numbers and see a company's growth rate BEFORE signing on? If you are the best in the game, you can pick the right startups at the early-mid stage when they need a CFO. No exit is the same. Some can be brought in the first two years, and others can take much longer. If you know the numbers, you should be able to figure when and for how much the upside is. Either way, having access to the numbers can be fairly enlightening, and you get to hear about news earlier than most employees.

Sample Financial Analyst Career Description

Let's take a look at a sample description of a financial analyst role. This role is much more math- and analytics-heavy than the other gateways. I've also **bolded** a few characteristics and soft skills that you'll need able to develop. A trend that I'm seeing from the other gateways that is happening here as well is a focus on leadership, communication, and getting things done!

Responsibilities

» Own major pieces of the company's financial modeling, forecasting, and board reporting process

» **Collaborate with finance leadership** team to set and monitor total company budget and impacts to the PandL

» Lead cross-functional teams to **drive initiatives** such as: optimizing financial systems, creating tools and/or improving operating processes, headcount tracking and budgeting
» Become lead business partner and **manage portion of General and Administrative budget**
» Analyze short-term and long-term financial implications, and provide support for decision-making
» **Build strong relationships** with business partners across the Finance and Strategy and Accounting leadership to maintain and develop best-in-class key operational and financial metrics
» Manage the monthly/quarterly budget versus actual variance analysis; participate in the monthly forecasting, planning, and accounting close activities
» Strong understanding of accounting and the ability to draw upon knowledge in order to produce accurate forecasts; work closely with Controllers group to ensure accuracy and completeness during period-end close
» Support monthly/quarterly standardized processes to enable timely, disciplined execution of internal and external reporting deliverables

Requirements

» 1–3 years experience in FPandA, accounting, banking, consulting, or similar field. Technology industry and SaaS experience a plus
» Proficiency in **financial modeling** (Excel) and **developing executive-level presentations** (PowerPoint)
» **Strong attention to detail with the ability to identify and quickly resolve data issues/needs**
» Positive attitude and collaborative spirit; enjoy working and problem solving with cross-functional partners

THINK BIG AND DON'T LISTEN TO PEOPLE WHO TELL YOU IT CAN'T BE DONE. LIFE'S TOO SHORT TO THINK SMALL.

— TIM FERRISS

» Strong drive for data integrity, process definition, and organizational alignment

To help give you an inside look into corporate technology finance. I've found two folks to with very different origin stories share their wisdom. Jason was a biology major at The University of Irvine California with no prior interest in finance. Angelina studied finance at Penn State and was laser focused on pursuing internships that led her to work for Goldman Sachs before starting her career through the Cisco finance program. I met both of them while working at the Cisco Umbrella (formerly known as OpenDNS). Cisco is a HUGE technology company with over 71,000 employees. Cisco requires a lot of people to help with the financial side of the business, ranging from accounting payroll to financial analysis and planning to reporting earnings to Wall Street.

Jason's Take on Tech Finance

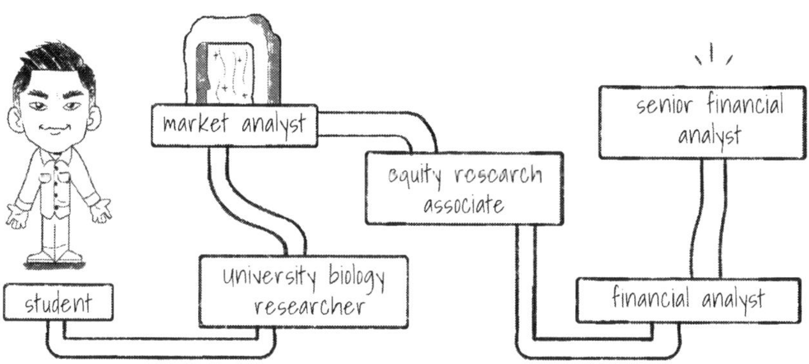

Irwin: How did you get into finance considering that you initially were a biology major?

Jason: I went to school at UC Irvine as a bio major. In bio, I was expected to take a lot of different science courses and a lot of math-based courses. In those courses, I definitely excelled and enjoyed running the numbers and working with a lot of different experiments, which is where I was able to do data analyses.

In my time in UC Irvine I also did research in a lab. To make sure our experiments were valid, there was a lot of math behind it. I liked the math, so at the end of my tenure at UC Irvine, I started to apply to more financial-related jobs to pivot my career and see where I would end up.

That's how I got to the Bay Area at a role with a small real estate investment company in Oakland. That company was only eight to ten people. It was great visibility. We worked with a lot of different investors.

It was mostly operations-based. I gained a lot of experience working with contractors and even our investors, sending out forms with our accounting team. It was a pretty good place to start.

Irwin: Where did you see that job posting?

Jason: From the college job board.

I worked at that real estate investment company for about 18 months. In that time, I learned a lot. Basically, I had no finance background. I learned how a ledger works, accounts receivable, everything about traditional finance.

From that, I wanted to learn more of a very traditional finance role. I went into equity research. In that role, I was covering banks in the Pacific

Northwest. During that time, I definitely got a really good basis of finance. I learned how to build models, how to make sure they're as accurate as possible.

I worked with a lot of different clients as well, calling from our New York office, and it was a great experience. I had to take four Series exams, Series 7, 63, 86, and 87, all within six months, which really pushed me to learn as much as I could.

That was one of those roles where I was up at 4:00 am and out when I was done. It was a really good grueling course to understand finance, and it was the hardest two years of my life, but I'll never regret it.

Irwin: Why did they make you take all those Series? How would you explain a Series 7 exam, in a nutshell?

Jason: The different licenses allow you to work in a brokerage firm. You take the Series 7 exam, so you can technically sell products to a client. Because I was in equity research, I needed Series 86 and 87, which are targeted only for research analysts because I needed to be able to comply with different things, such as, ethics.

Irwin: So, you took that exam, you worked for equity research, and then what happened after that?

Jason: In my office we had other analysts who were covering tech companies, for example, semiconductors and other technology companies.

I started looking to other roles more similar to equity research where it's really heavy with modeling. That's where I ended up with my first Cisco role, which was in corporate development plans.

In that role that's a very specialized team. That's a team that doesn't exist in all tech companies; it's mostly in large companies. Cisco is very acquisitive; we've done more than 200 acquisitions total.

Irwin: For those jumping into this with no technology background: Cisco's core business originally was networking. Then they expanded into multiple product lines, such as helping with data center products, security, and helping with collaboration tools (plus much more).

Jason: It was really interesting because I got to see every segment that Cisco covers and every acquisition or potential acquisition that was coming in. We looked at it from a financial, as well as an operational, standpoint. We worked with all the different teams, engineering, HR, and sales teams, to make sure that our model made sense and that it would make sense to buy this company. We were that front gate. It was an interesting role. I got to see a lot of different investments.

Irwin: What happened next?

Jason: Finance is huge. There's a lot of different sectors. I was in equity research, which is very different from the Cisco world. It was really interesting, but I was building models and doing projections for things that I never had a hand in. That's when I transitioned into doing financial planning and analysis (FPandA), working with budgeting and forecasting for OpenDNS as well as a different segment of the security business.

It's been really interesting. It's probably one of the most popular roles in this kind of finance world because every company needs to have an FPandA person who budgets, forecasts, and basically makes sure that they're spending the correct amount of money, avoiding overspending or underspending, and such.

Irwin: What have you noticed by being at a tech company versus a non-tech company?

Jason: From content, I mean, I was working with banks before, and that's very black and white. Things all operate in a very similar manner, so it was almost like a ritual, the same thing over and over again.

But once I got into tech, every business is very different. Basically, all tech companies run in a very special way. So that keeps the job very different. Keeps it more interesting. I'm not just running the business; I'm scaling a business. I might be downsizing parts of the business, and it's more interesting in that sense.

Irwin: What are the different ways in which people can break into the finance side of technology? What advice would you give someone, and if you had to do it all over again, at a new college, all over again, what would you do differently?

Jason: Honestly, although I took a very unique route, I think my science background was what made me stand out. I took a lot of those courses that basically make sure you have a really strong background in math.

I think no matter what path you take, there's a reason you took that path.

So there's always a way to come back and get to the place you want to be.

Irwin: Any advice for a person looking to break into tech via finance?

GOLDEN NUGGET

Jason: Sharpen your math skills. You also need to be able to understand SQL or Python and coding language. That's something you can basically learn online yourself, and that has been a huge requirement for a lot of new roles I am seeing now. You need to learn those things, so you can do data analytics. That's a key requirement that will probably be really big in the next couple years.

Irwin: What are the trends you're seeing and what are the skill sets? What are some of the things that you're seeing that are unique to your side of the business?

Jason: Coming back to basically understanding SQL, for example. With SQL software you can go into your database and type in a query, and it pulls it out. You can have very specific parameters for the data you're trying to pull out from your source. It's a big part of being able to extract the data you want and do analysis that makes sense.

Other than just budgeting and forecasting, which is very plain and more black and white, this is more of the strategic finance part of the role. It allows you to look at different deals from Salesforce.com and extract the data to make sure that you'll do different analyses, such as discounting customer accounts. It's very limitless in terms of what you want to pull.

Irwin: What are the different gateways or entry-level jobs into a technology finance team?

Jason: It usually starts at an analyst role. There's the very traditional route of FPandA, where you're doing budget forecasting for a corporation. There's also the banking route, which is very different. That's where I kind of started, being in equity research. In that role, that's called an "associate in equity research" or an "analyst in investment banking."

Irwin: What's the difference between a banking lifestyle and a tech finance lifestyle?

Jason: Banking lifestyle means you're basically just working nonstop. You have your phone with you and it's attached to you and you need to always respond.

Irwin: And did you make more money, in theory?

Jason: Yes. You definitely start with more, but I think it's one of those roles that people take on for about two years, and then they move on because they burn out. However, the experience, I think, is invaluable. It's a good route to take. Tech role, even though in finance, is more lax. It's nice. It's a good balance. It's a really good balance.

I've gotten to live both sides. I get to enjoy more of my hobbies and personal life now that I'm in tech finance versus before.

Irwin: What do you believe is the ideal profile? What are some qualities, if you were a hiring manager, that you would want to look for? From a person either coming in from banking or a graduate?

Jason: We look for someone who has some of a finance background. I was an outlier and definitely kind of slipped through the cracks that way. Definitely a finance background. If you're not in the program, we usually look for one or two years at a company where you at least have to understand profit and loss.

I think for all the other roles I've been in and when I've seen my hiring managers hiring other people, we've homed in on attitude. A thirst for knowledge has been really critical. I interviewed Angelina before she came in, and that's basically what we all agree upon was her very big attitude on wanting to learn more. And if she didn't have a skill, she was willing to take it on.

Angelina's Take on Tech Finance

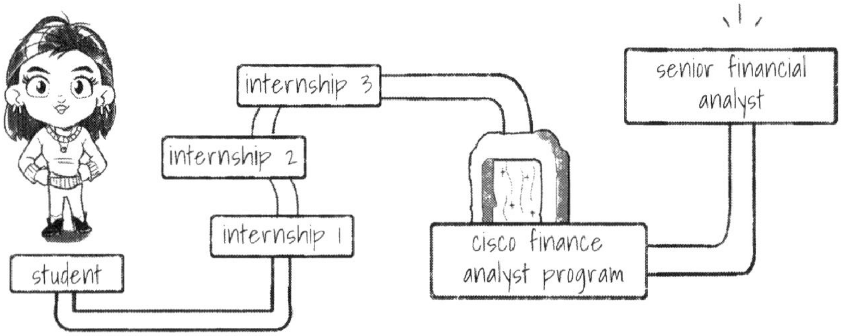

Angelina will share her story of getting to her dream job at Cisco. She has had a really cool career. Her story shares a lot of insights on how to break into finance, taking a rather unique route. She was able to leverage her internship experience into her dream career. Her interview highlights what going beyond the theoretical advice you receive in classes can do for you.

Irwin: Angelina, you are part of the finance team at Cisco. What's your story? How'd you break into finance? How'd you get into Cisco?

Angelina: I was born and raised in the Philadelphia area. I graduated from Penn State in 2015 with a degree in international business and finance. So that's definitely what piqued my interest, studying finance in my business classes. I had a variety of internships with a consumer packaged goods company, an investment bank, and then I accepted a full-time offer at Cisco upon graduation. I actually started off in their rotational program, which is two years long with four different rotations into different areas of Cisco finance. Then, after graduation, I joined Irwin here at Cisco Umbrella. I've seen a lot of different areas of finance with different companies and even within Cisco, different areas of corporate finance.

Irwin: It sounds like you are a professional intern breaker. You broke into an investment bank. Can you share a little bit about that?

Angelina: One of the great things about going to Penn State is they have a huge alumni network. A lot of corporate companies come there because an alum will graduate and then they'll bring their company back. I think the journey actually started my sophomore year. I met the consumer packaged goods company at a career fair, and they actually offered me an eight-month long co-op, which meant I had to take a semester off of school and move to Hershey, Pennsylvania. I was really hesitant about doing that as a sophomore, but it actually paid off because when I spoke to the investment bank the next fall at the career fair, I was the only junior that came in and had eight months of finance experience.

Most people in finance get their first internship the summer after their junior year. So being a junior and already having eight months' finance experience is what I think really opened the door for me. It was taking the risk and taking the semester off of school and really jumping in.

Irwin: How did you know to get an internship right out of the gate?

Angelina: I went to the career fair just to start practicing speaking with people.

Irwin: Freshman year?

Angelina: Yes, freshman year. The more corporations and businesses you can put yourself in front of, just to get the hang of talking to them and feeling more relaxed, it will set the stage for your junior year, which is when you get your most important internship. So putting myself out there, they signed me up for an interview. I thought they were just being nice, giving me the chance to interview. The interviews wound up going really well, and next thing I knew, I had the offer, so I decided to take it. It worked out.

Irwin: Sounds like you were super excited.

Angelina: Yes, I was really excited. It's a great company to work for. I've always worked at very large companies. I mean, all of them are Fortune 500, and then coming to Cisco Umbrella, it's still a pretty large company. But it's a little bit smaller than what I'm used to.

Irwin: What was it like working on Wall Street?

Angelina: Wall Street was intense. It's definitely a little bit different from what I'm doing right now. It's a different vibe. New York, I really think that's the finance capital in the United States. There are secondary capitals, which in my opinion are Chicago and San Francisco, but it was a great experience. The role I was doing was definitely not what I was passionate about, so I decided to keep my options open for senior year, and that's why it really interested me about Cisco's rotational program. Because finance is so broad, I wasn't sure what to do, so having the opportunity to rotate onto four different teams was an amazing experience.

Irwin: Can you give us an overview of that process? Including the application phase?

Angelina: It's about rotating throughout finance and really building up your core competencies and understanding the business to make yourself into a good leader. TAs far as the interview process goes, it was pretty basic: a 30-minute phone screening just to make sure you're a normal human being, and you can speak a little bit more than your resume. Then I had four back-to-back interviews, all over WebEX, which is a phone video conferencing system. Then I was extended the job offer.

It went pretty smoothly. I didn't have to travel. I think nowadays they have a super day where they bring everyone onto the campus for an interview, but I just had the four back-to-back interviews. Even after the program, finance employees are encouraged to rotate every two to three years, depending on what level they're at, just to continue building their expertise.

Irwin: If you were a college student graduating today with a financial degree and you were not so upfront on internships, what would you do?

Angelina: I really think it's about knowing the markets. I think going through the interview experience is important because what you learn in your college classes is very theoretical finance. It's really important having some real-world insights to speak towards because in classes, you are just learning a lot about accounting and different financial equations. Understanding real-world applications and developing a real passion for the market are important. In finance I can't catch up on the news in one day. I need to be able to speak to a trend that's happening over time. Something that's really important is just reading the Wall Street Journal, Bloomberg News, or CNBC every day because you can't catch up over one day. You really need a time point to compare it to, so following stories over time is really important and evaluating that against the market.

Irwin: What are the additional things that a student can do or someone out of college can do to differentiate themselves?

Angelina: The more you can get into data analytics involving finance is really important. We are dealing with very, very large data sets every day to complete our analysis. Anyone can complete an analysis, but also be able to take it one step further and actually understand how we get this data and drill down into it. A lot of that actually involves coding. So something I've been learning since joining the OpenDNS team is SQL coding. I wish I knew more languages and the issue is, they are a language, and it takes time to build them up. So the earlier you can start learning these languages, the better.

Irwin: So you're taking SQL classes?

Angelina: Yes, and people on my team use SQL, so I still need to rely on some team members to pull data for me. The industry in general is going to data. Everyone needs to know how to pick data, and that's where a lot of the value is, in data. So understanding how to pull that data and manipulate it is crucial. A big aspect of that is coding. So the more technical you can make your background, I think the better.

Irwin: For those that don't know, how would you describe SQL?

Angelina: It's a language used to drill into a database to pull out information you would like.

Irwin: You've done Wall Street. You've done Cisco. Now you're at a startup within a company.

What do you notice are the differences between a startup and a big tech company? And then what do you notice overall between a tech company and a non-tech company?

Irwin: I'm gonna start backwards. The big difference between leaving banking and going to technology is, in banking, everyone works in finance, everyone has a very similar mentality, so I thought. When you make the transition to working in finance at a technology company, you're more of the minority, so you get to interact with people who think in a very different way than you. Not only that, but several groups of people—engineering, sales, marketing. Being able to understand how those people think and get new ideas I thought was a lot more exciting than just a bunch of finance people in a room together, sharing a similar mindset.

About going from a larger company, Cisco, to Cisco Umbrella, which is more of a startup culture—a main difference for me is just how much smaller my team is. I went from a team of several hundred people at Cisco to only five. So it's very small. I feel like you have the opportunity to get more of a bird's eye view of a company. Right now I report to the head of finance. You don't really get that opportunity at a large company. Being able to report to the CFO here, I think, is really exciting. Getting more of the bird's eye view and having more of an impact. I feel like I can see the impact of some of the projects I've worked on more quickly. Things do happen a lot more quickly here, compared to Cisco. I feel like there's less red tape.

*

Jason and Angelina's perspectives are powerful for those going into tech via the finance route. While I've had a lot of friends take the iBanking and consulting routes, not everyone is made for that life. And before burning out or quitting, it's worth sharing another alternative: breaking into tech through the finance gateway!

GOLDEN NUGGET

Here are a couple of key points to help you differentiate yourself:

- » Start as early as you can! Remember Angelina's freshman head start leading to internships
- » Study the market and trends to stay relevant and smart.
- » Pick up skills like SQL, Excel, and data analytics type of studies.

When I was graduating university it seemed that the finance industry was hiring the majority of graduates. What I've noticed is that currently, more and more graduates are looking at doing startups versus going to work for big banks. Folks who are in the finance industry are joining the finance side of startup businesses. It's a beautiful transition. As you heard from Angelica and Jason, there are pros and cons of going into banking or doing the technology route. The key idea here is there is demand and choices for folks in the finance industry today.

If you ever wanted to know what skills you need to bring a product from idea to launch and hire and manage the team behind that, then that's what the next chapter addresses—the product management gateway!

CHAPTER 7

THE PRODUCT MANAGER GATEWAY

Have you ever wondered how applications, features, and products are launched? As the team is building a product, and it becomes ever more complex, there is a team member whose responsibility involves taking in the direction of what the product leader wants done and communicating and organizing the team's action items to ensure tasks are distributed and timelines are met or adjusted—welcome to the world of product management (PM). Some say it's a lot like being a CEO of an application while others say it's more of a manager of processes and people. There's some truth in both.

The product manager is a great role to be in tech. Marissa Mayer started as a PM at Google before eventually becoming a CEO of Yahoo. As a PM, you get to learn the ins and outs of what it takes to launch and optimize a product to get the most engagement possible—all without having to know how to code.

A good way to understand the PM role is to think of building a house. The general contractor leads the build. The general contractor is responsible for prioritizing and organizing all the subcontractors' duties (foundation, plumbing, electrical, framing, etc.) to ensure everything gets built according to the architect's plans, on time, and on budget. The PM is kind of like a general contractor, but for technology.

This gateway is very much a chicken-or-an-egg dilemma, as in "Which comes first?" A lot of well-established tech companies rarely hire PMs without prior experience, but it does happen. After doing much research, I highly recommend checking out ProductManagementHQ.com and a book called "Cracking the PM Interview." The website was started with the goal to streamline the process and answers on what it takes to become a PM.

GOLDEN NUGGET

Shoutout to Breaking Into Startups podcast team as I found our about Product Management HQ from them.

If you like the interview format in this book. Check out their podcast as it's full of other interviews across the Tech community!

I've learned that the best way to land a job in the PM space is to get educated, meet people in the space, and see what characteristics are wanted in a PM role. After that, go off on your own to get hired and then product manage your first solution (i.e., product), so you can have that experience to show off to your next company. Product Management HQ has a great online course and community to support you in the pursuit of going from zero to PM.

Gateways

- Associate/junior product manager ($67k–$123k)
- Ex-consultant or some type of finance experience ($61k-$149k)

Career Paths

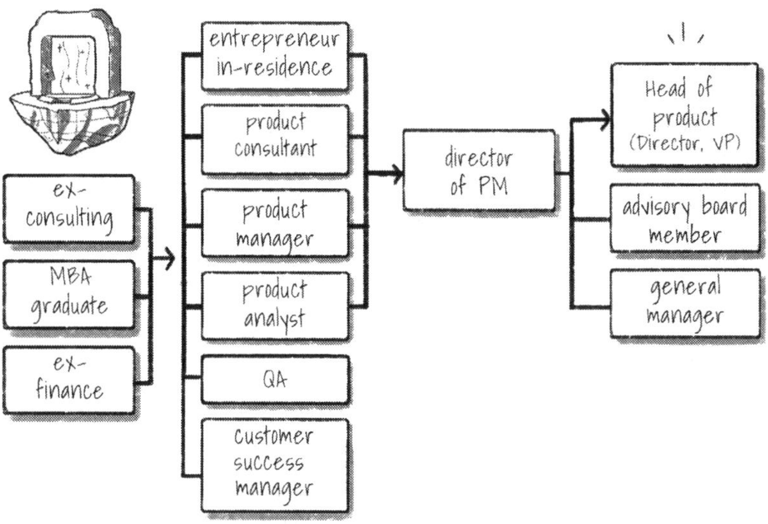

Skills

- Cross-functional teamwork and leadership skills
- Ability to make tough decisions and negotiate disagreements
- Prioritization, process, and roadmap management skills
- Analytical and metric-based estimate or projection skills

Sample Product Management Career Description

In this example, notice how the PM is expected to own the performance of the product. Not only do you need to drive the roadmap and development of the product, there is also an element of analyzing data to make decisions that will improve the financial results. I've bolded a few key points for you to pay special attention to too.

> As an associate product manager, you'll manage the product lifecycle for core games from initial concept through development to release. You'll own and manage a live game played by millions of people around the world!

What You'll Do

- Own the **analysis, product roadmap, and merchandising** of virtual goods
- Be responsible for driving **financial results**
- **Coordinate and collaborate** with multiple teams, resources, and timelines—engineering, art, UI, user acquisition, and marketing—to create engaging and profitable content
- **Write clear and concise specs**, and both wireframe and design new features
- **Analyze key success metrics** and **rigorously test** and **optimize for them**
- Work in an entrepreneurial learning environment with a world-class team of engineers and product managers

What You Bring to the Product Management Team

At least 2 years of experience in:

- Strategy/management **consulting or finance** (banking, PE, or VC)
- Online/mobile product experience with a technology company
- Experience in **quantitative analysis** and ability to **refine hypotheses/models** by digging into user data

» Strong design mentality and a familiarity and interest in mid-core/core games
» Most importantly, you should love the idea of creating highly engaging mobile content that will be seen (and played) by millions!
» Bachelor's degree required

Extra Gems for

» Previous startup, social networking, engineering/software development experience
» Degree in a quantitative discipline (CS, engineering, math, econ, etc.)

Luke's Take on Project Management

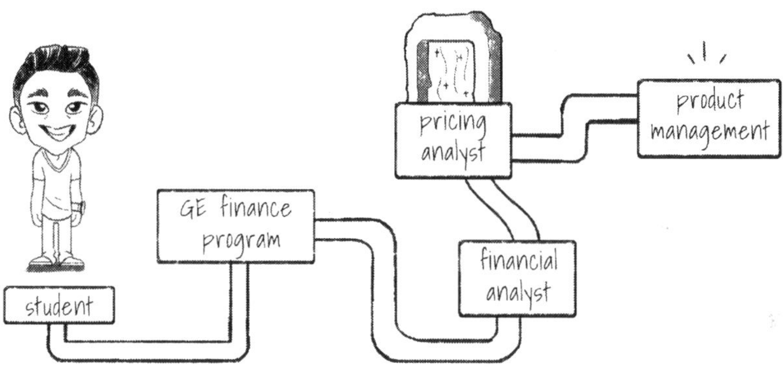

For the interview I reached out to my old highschool friend Luke. Luke studied business at The University of Carolina and found his way into PM. He's currently a senior PM at AppAnnie.

Irwin: I wanted to share a perspective of someone who broke into product management, and what's great about Luke's background is, he didn't have a technical background.

A GREAT PRODUCT MANAGER HAS THE BRAIN OF AN ENGINEER, THE HEART OF A DESIGNER, AND THE SPEECH OF A DIPLOMAT.

— DEEP NISHAR

But before I jump into it, would you share your story?

Luke: I got a degree in just regular BSBA, so Bachelor of Science in Business Administration at North Carolina Chapel Hill. And I came out of that with a lot of opportunities in consulting and finance, whether that's investment finance or corporate finance. I took the corporate finance route. Went to GE and kind of learned a lot about accounting, strategy, and analysis.

Irwin: How did you actually get your first job into finance?

Luke: On campus at Carolina, the business school had a really good, robust program to bring in recruiters to help you as a major find opportunities and roles in the industry.

I grew up in North Carolina and went to school in North Carolina, so most of the opportunities were focused more on East Coast type of companies. So, you had a lot of people like the General Electrics of the world, you had your Accentures, your big banks, you had your big consulting firms. But you didn't see any Googles out there. You didn't see any LinkedIns, Microsofts, or any of those sorts. So, it was really interesting, because the tech world seemed very far and away. For me, personally, when I was leaving college, I knew very little about tech. I didn't even know what product management was, and really all the opportunities in front of me were more focused towards finance and consulting.

Irwin: And then, where did this job take you, and what were you doing at GE finance?

Luke: I spent about two years in GE's financial management program (FMP), which is a mixture of four different jobs across two years. Learned a good bit, like I was saying earlier, about the fundamentals of accounting, and the fundamentals of running a business from a financial perspective. It was really helpful because it helped me get really comfortable with numbers and with business strategy.

Those were two things I took with me throughout the rest of my career, and they really helped me get into product management because I was very familiar with analysis and numbers. I understood how those numbers impact a business.

The third thing that really helped with GE was, I lucked into a couple roles where I worked a lot with the business systems. So, how do we improve the systems that we have at GE to be able to better service our needs as a finance organization? And that gave me a taste of how systems interact with businesses and how certain tech can help you as an owner of a function do your job better. Those were some of the fundamental principles that actually helped me contextualize what I did in the past and apply that learning to my later work in product management.

Irwin: So, you built two different skill sets, an analytical skill set and then kind of seeing how it all played together from a technology and how it interfaced with the business. How did you break into tech from there?

Luke: I did about two years in the FMP at GE. After that, I did a couple more years at GE doing an assortment of finance and pricing-related roles. I actually broke into a software-related pricing role and worked in Boston for a few years.

From there, I dabbled a little bit in technology. I knew I had a passion for software and technology, and I wanted to get into that field. So, I did some things myself to learn a more about how technology works—a bunch of free, online open courses to understand some of the basic fundamentals of computer science and how operating systems, applications, and the like all work together, at least on the high level.

Some other things that I learned that helped me out later was learning how data works. How data flows through systems. Very fundamental concepts like relational databases and SQL, which is a very fundamental building block of data.

Those were some of the skills that I picked up while I was still at GE. Then at some point at Boston, I actually created an AngelList profile, and through AngelList, I was contacted by a startup out in San Francisco.

Irwin: Could you explain what AngelList is?

Luke: AngelList—think of it as like LinkedIn, but primarily catered towards people at the startup stage of the world. You can build up your own personal profile, but what's really interesting about AngelList is you can get a wide array of different companies, particularly at the startup end of the market.

GOLDEN NUGGET

I knew that I wanted to do something in technology, and I wanted to do a startup, and I've been kind of eyeing going out to San Francisco for several years at that point, so when the opportunity came along, I got contacted by a pricing startup. I jumped on the opportunity and came out to San Francisco.

What was really fortunate about that experience was, because it was a startup, they were willing to take a chance on somebody who didn't have a lot of traditional product experience. They were just looking for somebody who understood the space and understood pricing, and seemed pretty smart and could put things together. So, they brought me into the fold and hired me, even though I didn't have any product experience, and gave me the room and the cushion to learn product as I was on the job, but also I applied into the role some of the industry expertise that I had built.

Now when I hire product managers, I actually look for those two dynamics. How well does this person understand the fundamentals of being a product manager? And then, how well does this person understand the industry and the vertical? The ideal candidate is somebody who has both skill sets, who is an expert in the subject that we're doing and who also knows how to be a good product manager.

Irwin: Could you explain what a product manager is? What makes a great product manager?

Luke: One big clarification to make when we're talking about product management is, at least in the experiences that I've had so far, it's product management in, what I like to call the 2010's and beyond. So, this is your big, commercial, new age technology companies. The reason I make that delineation is before 2010, product managers were much more technical in nature. You're talking about former engineers, people with very deep technical backgrounds, because these people would have to write very detailed, technical specifications so that engineers would build products in a very precise, right way.

A lot of these product managers, and kind of what I think of as the older school, were significantly more technical, and the technical chops were emphasized a lot more; whereas today, when I look at product management and the types of people that go into product management and the skill sets that hire for, they're definitely looking for people who are a bit more strategic thinkers, who are a little bit more a jack of all trades. These are people who can speak the customer language, who can be very customer- or user-focused, depending on what your business style is, and also people who can take that and sit between the customer and the engineers.

What's interesting is, engineers have become a little bit more customer-focused now too, so the actual technical chops of a product manager, at least in this generation, are a little less emphasized; the emphasis now is on how well this person can think about all the different dynamics: is this person a logical thinker? Are they a strategic thinker? And can they communicate well with all the stakeholders and manage cross-functionally?

Going back to your previous question, some of the skills I look for when I'm vetting a product manager is, number one, whether the person is a clear thinker: are they a good problem solver? Can I put a problem in front of them, and can they dissect it down into who are we solving the problem for? Why are we solving it? And they need to be able to break it down into all the pieces that may be involved with solving the problem.

A good example: how do I improve some sort of product for an existing base of users? They're going to have to really consider what things the users need. And then they're going to break it down into all the different options that we could build. How do I distill the MVP (Minimum Viable Product), the most important thing that we should ship out first, to test whether or not we have something valuable or not? How do I figure out all of the additional features that can come after that, and then create that product backlog? I look for a very clear, logical progression that says, this is the thing that's going to deliver the most value, and then here's how we're going to build on top of it. All the while, the person should be very eloquent about it to illustrate readily how the story is going to work.

Irwin: Let me know if I'm articulating this correctly: part of the job of a product manager is to figure out what's the need of a feature, a function of a customer that it's meant to serve, and to take direction from a VP of product, to go build this or go do research into this.

Luke: Yes.

Irwin: And then, finding and then roadmapping it. When you said "backlog," is it looking backwards, meaning—to get a product like this, what are the different steps that need to happen before that?

Luke: Exactly.

Irwin: And then, determining which stakeholders we are talking about. Like what team members do you have to work with to drive the completion of that product?

Luke: Yes. It starts with understanding, as in, "We have a problem we want to solve, and we have a good understanding of who we want to solve it for." That's the very basic fundamental of building great products. Have a deep empathy for your user or customer, and understand how you're solving that problem.

From there, you might have a roadmap or a backlog of features that you want to build. Next, it really comes down to what I think of as the next level of product management. So, there's the, is this person a strategic thinker? Is this person a critical thinker, and can they create this backlog? And then it comes down to, how well do they work with their counterparts across the organization? Primarily, it's going to be engineering. It's how you work with your engineers to communicate this backlog, to communicate this vision, so your engineers have a good understanding of what it is you're trying to build. Then, as a product manager, can you give them enough specifics and speak the language that engineers speak so that they can build it in a fairly efficient manner?

That' involves being able to break things down into logical components. Not necessarily saying something like, "I want you to build an email form," but really getting into more of the specifics. "I want an email form to achieve this goal, and then when users do X, I want the email form to be able to respond with Y." Giving some give and take, and understanding, if I'm a programmer and I'm writing code, how might I approach that? What are some of the flows that I really want to be able to focus on?

Going forward from engineering, there are a couple other departments that you definitely need to work really well with. Depending on the type of company, design is very important. Particularly in consumer products today, design is a integral element.

Irwin: For those who don't know, design is the user interface, right?

Luke: Design is the team that's going to build that user experience (UX), that user interface to make sure that when people are interacting with their product, they understand how to use it, and it's intuitive enough. And also, for a lot of these consumer products, that they're delighted to use it. You think about your user experience when you're on Instagram, think about how beautiful, simplistic, and easy it is to use. That's a culmination of hours and years of design work and design research. Design is a very important partner when it comes to product, and being able to convey, once again, your strategy and direction to the design team is crucial.

The final team that most product managers will probably have to work with, outside of sales and customer success and marketing, is data science. As the industry becomes more and more data heavy, being able to work with data science and being able to speak in a language that data scientists understand is getting more and more important. That could be something as simple as saying, "I want to be able to solve some sort of data science problem to improve the outputs in my existing product," or "I want to package this data science solution, maybe it's some sort of machine-learning algorithm, some sort of statistic output, and give that to my users." It could come in a lot of different forms, but being able to be fluent in data and data science is becoming more and more important as product managers are finding themselves working more frequently with data science.

Irwin: What advice would you give someone who would like to pursue looking into this career?

Luke: One big piece of advice I would give to somebody thinking about becoming a product manager is doing that two-way assessment. On one hand, really understanding what a product manager's job looks like. What does their day-to-day look like? And I would say that exists for any kind of job or any kind of industry you want to get into, meaning, if I got into this job, what is my everyday going to look like? What are the types of skills that I need to have in order to be effective at this job? And what's the type of work I'm going to do? Am I excited to be talking to a lot of different stakeholders every day and working cross-functionally? Am I okay being somebody who's at the middle of many different departments driving an outcome, even though none of these other departments report to me? Am I really good at communicating complex thoughts in a concise manner? Those are some of the things that you might want to think about to really understand what a product manager's job looks like.

The second part I would actually think about is the other side of the self-assessment, which is, what are my skills? What am I good at? And particularly, if I'm already in a job, where is my career trajectory taking me? Maybe you're an analyst right now and you're very fluent with data. Leverage that as a strength. Maybe you get a product analyst role, so you can do analysis in a little more strategic light, and then that could be a gateway into product. Maybe you're in a customer success or a commercial-oriented role, so you have a really good tie with the customer, then you want to think, "How can I take my job and transition that to something that includes something more technical? Maybe I want to beef up my technical chops so I can learn how to speak better with engineers and talk in their language?"

That two-way street of understanding is critical, both (1) what makes a good product manager? And (2) what are my skills, and where am I in terms of my career?

In terms of entry points, I think of it in a couple different buckets. I think of it as, maybe you're in college right now, and you're thinking, "What do

I want for my first job? Maybe I want to end up in product management tomorrow. Maybe I want to end up in product management later on." There's a lot of entry-level product management programs in the Bay Area. A lot of the big companies, like Facebook, Google, and LinkedIn, will have these rotational programs that are about two years in length that teach you the ins and outs of being a product manager.

Those are pretty competitive and great to get into. If that's not an avenue that you're considering, another avenue that I've seen work really well with product managers is getting into management consulting. A lot of the skills from management consulting transfer over into product management really well. It's that strategic thinking, getting deep into problems, trying to find the crux of the problem, and then being able to communicate that in a really well-written fashion to all your different stakeholders.

The third one I would think about, if you're thinking about getting into product management, is something like private equity or some sort of banking background because, with that kind of background, people know that you're strategic enough, but you're also very detail-oriented, and you're very good with numbers. That helps you get an entry point into product.

Program/project management vs product management

One final one is program or project management. There is a lot of need for people who can be partners with product managers. Program or project managers are usually the ones that will make sure that the train stays on its tracks. If the product manager is defining what we're building and why we're building it, the project manager will make sure that the engineering teams and other teams involved will execute and keep the trains on the tracks. That's a really good opportunity because you understand how teams work and how product gets built, and then through exposure to product managers and product strategy, you can start building that intuition and the other skill sets that will be required to be an effective product manager.

Those are a couple different entry points in terms of the jobs that you could explore if you're really looking to transition into product. The second bucket of people I would consider is people who are already in industry. Maybe you're a few years into your career and you're thinking, "How do I transition my career into product today?" If you're fortunate enough to live in the Bay Area, there's a lot of good product schools in the area that teach the basics of product management, and they have a lot of good connections with existing tech companies to help you land that first product job.

If you're not in the Bay Area, what I would consider is one of two things. One is business school. I've actually seen a lot of people who go to top-tier business schools and then end up in a product role on the other side of business school. Usually, that means the person is quantitative and a good critical thinker. That kind of acumen will help you get your foot in the door for potential future product jobs.

If business school isn't the way that you're thinking, then you've got to be a little more crafty. Maybe get lucky like I did. There's two different ways I would think of here. One is to go to an existing tech company and do the job that's very similar to what you're doing today. Get your foot in the door, and make sure you're on that rocket ship. When you already have that foot in the door, and people know that you're a good worker, a critical thinker, and you have the acumen that's required to be a product manager, then you can look for a horizontal move into product.

> GOLDEN NUGGET
>
> And if you don't go to a big tech company, you could go find a small startup like I did, and those companies are typically willing to take a little bit more risk, and maybe hire somebody who doesn't have a technical or a product background to be that strategic thinker, to be that person who can pick up product as they go. Those are a bit riskier, but you can usually start picking up your product chops more quickly that way.

Irwin: I think it's worth mentioning that whenever they reached out to you from AngelList, it wasn't for PM, it was for analysis of pricing, right?

Luke: Yes. When I joined the startup here in San Francisco, I wasn't originally hired for a product title. I stumbled onto the product title after I arrived in Silicon Valley. I was brought in to be this pricing expert and the lead of all things. Basically be that subject matter expert for the company.

As a result, a lot of my subject matter expertise got passed into what we wanted to build into the product. A combination of that expertise and my own natural inclination and passion for building things that have impact created this fusion where I stumbled into product management. Also, I learned a little bit about it by reading a bunch of blog posts and talking to a lot of people who are existing product managers.

I found people who were product managers and asked, "Hey, you want to get coffee?" And then asked them about their experiences.

Irwin: That's an informational interview (More in chapter 17).

Luke: Yes, and I learned a lot about how other people do product and how other people have built their product organizations. When I was at that startup, they didn't have any real product experience there, so by learning, even the basics of Scrum and Agile, how engineering teams work, and then also talking to people who are already working in product, I picked up a lot of information in my first year to figure out what a product manager does and how I could be an effective product manager.

Irwin: How did you get these informational interviews? Did you just find people doing searches of product management and reach out to them and ask for coffee?

Luke: I got kind of lucky to be honest with you. Our founder would actually bring in product experts to talk about certain finer points about what we were working on in our startup. These people came from the network that was already existing in that company, maybe through investors or some of the founders. Also, friends who were in the industry who would give me warm introductions to people who they were working with.

It was really helpful. I got to have coffee with some really good people who, I would say, were three to four years ahead of me in that process. It was fresh enough for them where they could tell me a little bit about some of the pain points and some of the mistakes that they'd learned from over the last three years.

Irwin: Are there any questions that I should have asked you that I haven't asked you?

Luke: Maybe, what's my favorite part about being a product manager?

Irwin: Yes, tell us.

Luke: What I love about product management is, you go into work every day and you could get faced with any kind of problem, and the problems are so wide-ranging, and they're often very interesting. It could be something as exciting as, how do we break into this new market, and how do we quickly put together a product to attack this new market? So, something as strategic as market sizing, to something as frustrating as, I only have three engineers, but we really want to build these three things, and we just don't have the resources to do it. How can we come together with a creative solution to make something happen? To something as even more frustrating as, I've got 5 or 10 different stakeholders from across the business. Sales, marketing, engineering, design, and we're all trying to come together on a project. How do I get everybody on the same page, establish the same core vision, and be that one person to drive this product forward across all these disparate departments to bring them together for one unified vision.

They're all very different problems requiring different types of skill sets. I find that kind of thing very interesting and invigorating.

The Four Ways In

A quick recap—Luke summarized the four different ways people break into the PM space as the following:

1. Get education through training programs or courses like Product Management HQ.
2. Find a startup that is willing to take a chance on a first-time PM (Luke's way into tech).
3. Grow within a company through one of the many gateways in this book. Build relationships to jump into the PM space latterly.
4. Enter the role via a background either in consulting or banking.

If you ever wanted to be a build a product and didn't know how to code. This is the place for a non-techie to start. In fact, if you look at the career path of someone in PM: product consultant/speaker, entrepreneur-in-residence, internal lateral shifts, advisory board member, general manager, serial product scaler. I would say this gateway has a very interesting path.

But what if you want to understand what you are managing? If fact Marissa Mayer started as a developer first before becoming a PM. The good news if you don't have experience with computer science there are different ways to break into the developer side of the business. With the growing demand of technical talent there are more resources than ever available to learn how to code! In the next chapter we are going to cover the developer gateway.

CHAPTER 8

THE DEVELOPER GATEWAY

Imagine a world where you dream of an idea and see it come to life. You can see every detail and see how it works. You can create almost anything. You can type letters and numbers that can result in creating automation and ways to solve tasks that would typically take a very long time to do manually. And the only thing that is different is the magic happens by writing in a language that's foreign to most. With this skill you can build something from nothing.

When you think of developers, consider a young Mark Zuckerberg. He began using computers and writing software in middle school. One of his earliest projects was the development of "ZuckNet," which allowed all the computers between his family's house and his father's dental offices to communicate with each other. He basically created FB messenger for his father's business. Then at Harvard, he saw a need to help students make class selection decisions based on the reviews and choices of other students. He developed a program called CourseMatch to solve that. Just for kicks he developed a program that caused him a bit of trouble where he let students pick who's the best-looking person from a choice of photos. Then Harvard shut it down.

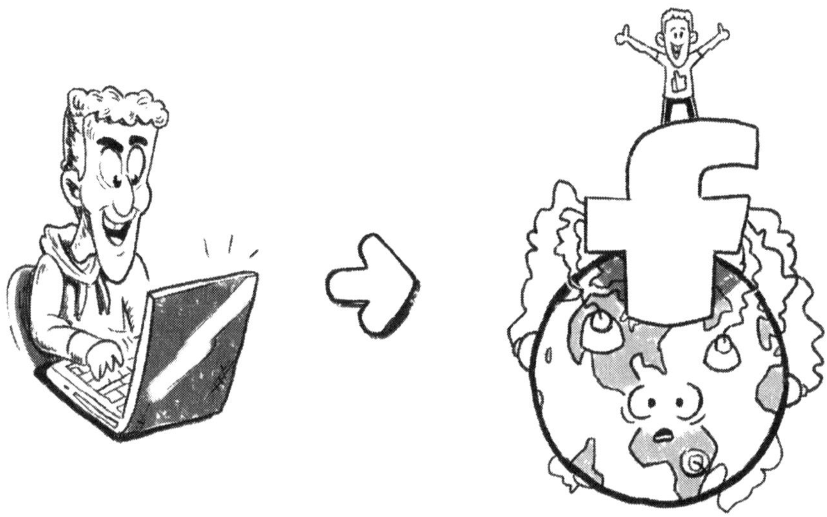

Eventually, he would develop his first version of Facebook in his Harvard dorm room. He had vision and the ability to create products from an idea. Now he's one of the most impactful and wealthy humans on Earth. His vision, and now mission statement, is to give people the power to build communities and bring the world closer together. This simple mission resulted in Facebook's creation and acquisition of new products and features like Instagram, Oculus, Whatsapp, and more. The amount of value the platform created in less than two decades has made his employees and investors extremely happy. In fact, the first 100 employees of Facebook have made more money from their equity than most business owners will ever earn in their entire lifetimes.

Most of the first 25 employees at a budding tech startup will fall under the developer gateway, as first you need a product before you can hire the teams mentioned in the previous chapters. Most people will shy away from this gateway as it requires a jump into the unknown to learn something from scratch. The great news is I have found related gateways that can help you break into the world of developers! If you can master these, you'll be sought after from every technology company in the world.

Gateways

- Quality assurance (QA) tester ($46k–$96k)
- Software engineer ($94k–$156k)
- Product managers ($94k–$169k)
- Learning coding: on your own, through courses, coding bootcamps, etc.

Career Paths:

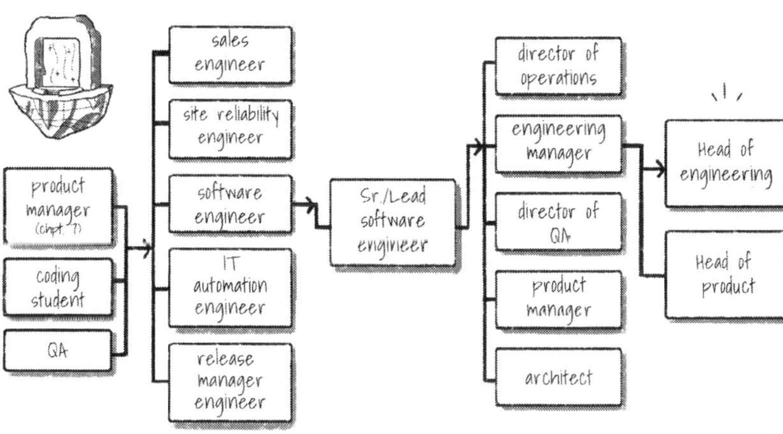

QA manager, software manager, product management, VP of engineering, chief technology officer (CTO)

Developers Simplified

- Quality Assurance (QA): the team who check for bugs and run tests to the code.
- UX team: the team that creates what you see on the product (the user interface), analyzes and configures how users are behaving with the product, and communicates their findings back to the business.
- Software engineer: the writers of the almighty code
- Product management: the manager guides the technical team and UX team in discovering and developing the right product for users.

Skills Needed

- QA: detail-oriented and able to follow instructions to check through software
- Software engineer: programing experience in 1 or many languages (Java, C++, etc,). You can either study on your own via online courses or get a degree in computer science (coding-focused) or engineering (hardware and software integration-focused) at an established college, which will take four years.

Sample QA Tester Career Description

Quality assurance or code debugging is one of best gateways into the developer gateway. As you'll notice in the QA career description included below, the QA job is to check the code that has been written by the developers and to run automated "black-box" testing tools that perform automated checks. If you have an eye for detail and are very process-oriented, this may be a place to start exploring. If you read the **bolded** points in the career description below, you'll see that there is definitely some self-learning needed to explore the QA methodology. You'll need to get a foundation before you can jump in.

> We're looking to expand our QA team by adding more talented individuals who will: ensure a high level of quality, quickly become a product expert, and advocate for user experience. While on the product team, you'll have a tremendous impact on our customers and will work directly with developers, product managers, and customer success. This will all be accomplished while embodying our core company values, which include delivering amazing experiences for our customers, being ambitious, and working as a team. Also you see here at the very bottom you see the company listing out their perks on the different benefits and leadership they have guiding the business.

A Day in the Life of a QA Tester . . .

» **Reviewing and analyzing** product requirements for **clarity and consistency** with existing features

» Creating and executing **black-box test plans**

» **Writing and verifying defects**

» **Finding clear reproduction steps** for issues reported by customers

» **Verifying builds** for quality prior to production deployment

Requirements, You Should Have . . .

» 1–6 years of software QA black-box experience

» Understanding of general **QA methodology**

» Excellent written and oral communication

» Methodical approach to **solving problems**

» Basic understanding of web-based applications

» **Extreme attention to detail**

WHO AM I? I DON'T WORK IN SILICON VALLEY. I DIDN'T GO TO COLLEGE. BUT, I'M CURIOUS. IT'S SOMETHING THAT CAN BE LEARNED. IT'S LIKE ANYTHING ELSE. IT'S A LANGUAGE (CODING).I THINK A LOT OF PEOPLE LET IT GET IN THE WAY .

— KARLIE KLOSS KODE WITH KLOSSY

Brownie Points ...

- » Understanding of CRM
- » Understanding of Agile development
- » Basic Selenium knowledge

QA as Your Way In

There are three ways in which you can learn coding: (1) through a four-year degree at a college, (2) through an onsite boot camp, or (2) teach yourself to code through books or online courses. My high school friend Len is a great example of someone who never studied coding at college and instead learned on his own.

Len went to NC State and studied business. As he was approaching graduation, he wasn't sure what to do. His friend pointed him to looking at quality assurance (QA). This experience was eye-opening in that Len argues that anyone who has the ability to follow line-by-line instructions and has an eye for detail and process can add value to the development team. He also explains that a lot of people get started in tech development in the software-testing (aka QA) space with little coding background; hence, QA is the developer gateway. Here's that interview.

Len's Take on the Developer Gateway

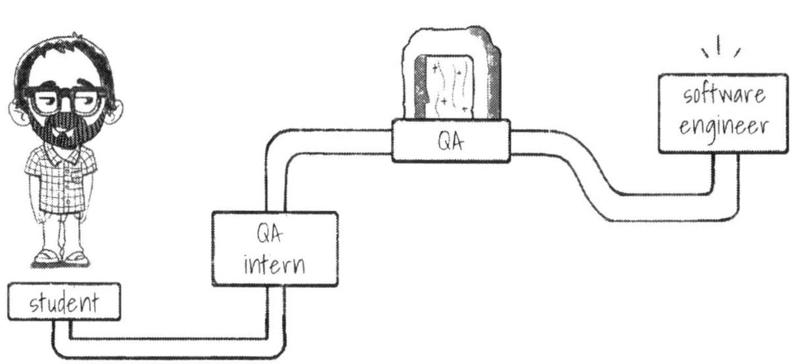

Irwin: Len and I went to high school together at the North Carolina School of Science and Math in Durham. We both graduated together, and then we went to NCSU. Neither of us has a technical background, yet we both (eventually) went into tech. I went into the sales route, and Len found his way into the technical route.

Irwin: I would love to hear your story and how you broke into the technology side of the tech business.

Len: In my senior year of college, I didn't actually have anything planned. I was going to be graduating with a bachelor's in business. I couldn't find an internship, so I asked a friend who recommended, "Why don't you join Fidelity Investments as a QA?" So I put in an application.

Irwin: What's a QA?

Len: QA stands for "quality assurance." And quality assurance is basically like a bug detective. When you use software, there are things that can go wrong, for example, the Blue Screen of Death on older machines. So, there are people looking for those errors.

If you're trying to make sure some software works, you are a quality assurance engineer. You could actually probably say, if you're editing someone's resume, if you're editing someone's document, if you're an editor, you could be a QA.

Irwin: Let's return to your buddy at Fidelity, who told you, "Check out this QA position." Did you have any prior technical experience?

Len: None. I did have a little bit of finance experience only because I was interested in the stock market. With this bit of interest in the stock market and a certification or two, the A+ and Network+ Certifications, I was able to say, "I'm really interested in what you guys are doing, and I need an internship." And they thought it was good enough. Which is basically what a lot of these companies are looking for in interns, right? A willingness to learn. They just want you to prove that you're trying to learn.

When I came on, I was an intern for about three months. Now, this was manual QA testing. So I was basically clicking around an application, making sure that the application was following the rules. Imagine you're testing your calculator, right? And you type 1 + 2, and the outcome is 4; that is a broken test. It was very tedious work.

I started realizing that software was probably a better solution for this problem than me clicking around. So I learned a little bit of code in my off hours and realized how important it was.

A lot of finance actually runs in Excel, and Excel has this cool language called Visual Basic for Applications (VBA). This stuff can really help you out. I was writing some Visual Basic for Application stuff in Excel, and it sped up my work a lot, enough that by the end of my internship, they decided to keep me on during my last semester of the school year. And then, that transitioned into an automation role. So, I started out at Fidelity as a manual QA, and then I went into QA automation.

That little bit of automation was enough to get me into where I needed to be. Then I reached out to you, Irwin, and you were able to get me into OKTA through their employee referral program.

Irwin: When people are looking to break into a company, a lot of technology companies have referral bonuses. So, when you, Len, reached out, I was like, "Yes. Absolutely. I'll get you in front of a hiring manager as fast as I can. Get that resume over to me." At the time, we were getting paid five grand per referral, so there's an incentive for people to refer candidates.

Len: You weren't in engineering, yourself, yet you helped me get into the engineering world. That is how I got my foot in the door.

Eventually, I put in a request for a position to start writing some software. It took a couple months, but they got me into a software engineering position. I've been in that software engineering position for about two years, and I've progressed now to a senior software engineer.

Irwin: What are the different ways that people progress from QA?

Len: Customer success, QA manager, and then a software engineer. If you decide to go the management route, that gives you a lot of flexibility to bounce into other management roles. So, my former manager at OKTA is now a senior engineering manager at Uber.

Irwin: There's two different types of people: there's the people in college, and then there's people who are working, but they want to break into tech. What piece of advice would you give them?

Len: I would recommend getting a small credential on your application. Something that has a little bit of tech in it.

Irwin: What would that look like?

Len: Take some Coursera courses on anything related to software. You could also do Code Academy. I took the A+ Certification and the Network+ Certification, which took a couple of weeks and is more of a traditional route. This was just enough that I could talk about technology in a meaningful way. The course was simple too. You pass the test at the end, and then you put it on your resume.

GOLDEN NUGGET

Len's point on adding certification and skills via courses will add more value to your career search than traditional degrees.

Irwin: What are the differences you see in starting with QA in a tech company versus other types of companies?

Len: In a tech-focused company, they are much more willing to dedicate resources to do automation. There's a lot of opportunity to learn how to do automation. Now that's not to say that we don't have manual QA. We do. We have some manual QA, but this usually is a stepping stone to learn automation. You have to move a bit faster. That means you learn faster, which means you grow faster, which means your career grows faster.

So, if you get into the QA space, I think you have a stable position. Manual QAs aren't going anywhere anytime soon. There's a lot of room to grow and a lot room to learn.

*

Len's interview was a pleasant surprise for those who can't code but want to be part of the developer team! And it's a great way to grow yourself to a software engineer like Len did.

The QA path is a great gateway job to get in the game and support the development team. If you want to skip the QA path and go straight to coding, lucky for you there are more free content, online and in-person training academies than ever before. There are even mobile games to help you learn the basics of code in a fun way, for example, lightbot.com.

After reviewing the ways numerous folks who broke into software engineering and seeing a personal friend of mine move from sales into a coding boot camp to become a product manager at Facebook, it's apparent that an immersive experience offers the fastest way to learn how to take yourself from a level 1 coder and land a tech job in as little as 16 weeks.

Here's the recipe for **accelerated** success:

"LEARN FROM SOMEONE WHO HAS ALREADY CHARTED THE PATH."

1. Start for FREE and see if coding is something you want to do:
   a. Freecode Camp, Treehouse, Code School, Khan Academy is dedicated to teaching courses for free including coding!
   b. Udacity is a great place to learn specific courses called nanodegrees. That's a great place check out as well.
2. After doing #1, then answer this question: do you want to pursue programing as your profession? YES or NO
3. If your response to #2 is YES, continue learning online or apply for a coding boot camp.
4. If you want to see my recommended resources go to irwinki.com

It's best to have some experience to ensure you can keep up with the boot camp program. Some of these boot camps have a 99% graduation-to-hire. These boot camps help address a huge problem in our business: there are a ton of startups popping up, and each one is competing for new talent, so these courses and camps help participants to focus on training skills that are meant for the technology business!

Udactiy is an an online academy where you can get nanodegrees (think very focused topics like virtual reality (VR), autonomous driving, self-driving car, AI, data analysis, machine learning, and many more). It's their approach which is so mind-blowing. There is a huge demand for skills in the areas listed. As it stands, the way most people have learned those skills is from the companies they're working at, meaning outside courses haven't been available. Udacity aims to train more people for these types of roles, so that way there will be more people with those skills that can join companies or form their own. They have partnered with the tech leaders, like Google, Facebook, IBM, to help build the curriculum and content for their nanodegrees.

The AI nanodegree course entails six months of part-time study and a fee of $1,600 (at time of writing). If you add a nanodegree to your toolkit, your market value will easily increase several tens of thousands of dollars (possibly millions if you happen to work for the next AI startup that blows up!).

Being part of the development side could lead you to different parts of the world, careers, companies, and adventures. In fact, there is so much need for this specialized talent that there are entire divisions (technical recruiting) within a company specialized in helping the founders and their companies find, retain, and train talent. That's the topic of the next chapter: the people (also know has human resources) and recruiting gateway.

CHAPTER 9

THE PEOPLE (HR) AND RECRUITING GATEWAY

Have you ever wondered who is responsible for helping the executive team build a team? Or what happens in the backend process after you have signed your awesome offer letter (see chapter 19)? What about when you first enter a business, who's the first person to greet you? Who makes that schedule you get for the onboarding activities that happen in your first few days of work? Who determines and organizes the corporate-required training? How does money start magically appear in your bank account? In this chapter, we address the part of a business that is often overlooked and is arguably the most important part—the people team!

The people team works directly with management in understanding strategically where the company is headed in the short and long terms, as well creating and enforcing a culture that aligns with a company's core mission and value statements. This team is also tasked with setting up compensation for all roles and benefits packages, like insurance, 401(k), and stock options.

The recruiting and training team is there to reach out and hire the right people and help with onboarding and continuous training. The people team helps investigate complaints and ensures that employment laws and regulations are followed and implemented to ensure a safe workplace. What about simple but significant items that should not be overlooked, like employee recognition, birthdays, weekly town hall announcements, and events? These tasks falls under this team.

A lot of times it is with someone on this team that's the first and last contact a person has with a company itself. If you are a people person and genuinely care about people and relationships, I'm excited for you as we dive into the people and recruiting gateway!

Career Paths:

» Recruiting path:

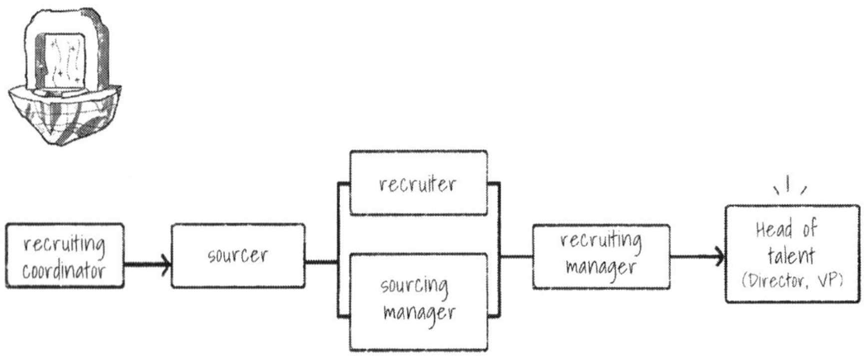

» HR path:

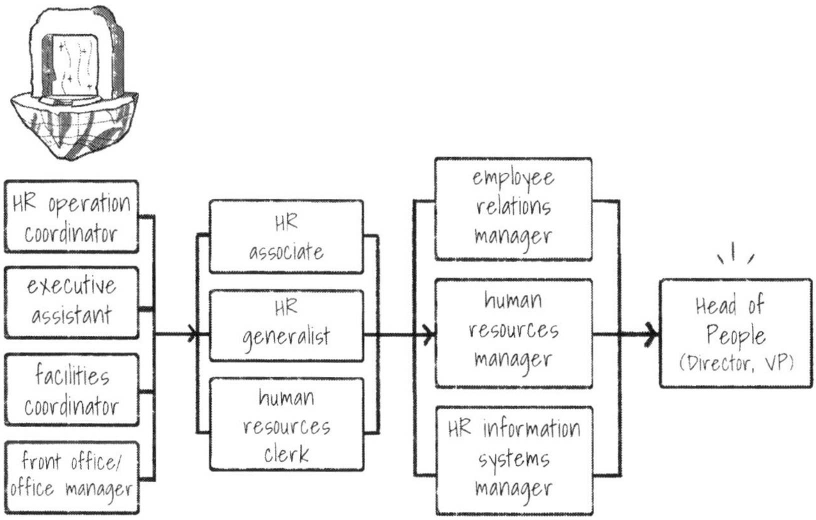

Gateways

» Recruiting coordinator ($36k–$67k)

» Training Assistant ($22-23/hour)

» HRIS Analyst (71k-113k)

» People (HR) coordinator ($39k–$61k)

» Benefits coordinator/assistant ($42k-$72k)

» Compensation analyst ($71-$115k)

» Front office/office manager ($34k–$71k)

» Facilities Coordinator ($44k-$107k)

» Executive Assistant ($61-$113k)

People and Human Resources Simplified

Here are their job descriptions for the different ways you can use HR as a gateway into a tech career:

- **HR generalists:** for the jack-of-all-trades and multi-tasker extraordinaire, HR generalist roles may be the fit for you. The HR generalist has a hand in all pieces of the HR machine, including benefits, onboarding, performance management, talent acquisition, and compliance. The HR generalist plays a vital role in ensuring that everything runs without a hitch.
- **Recruiting or talent acquisition specialist:** this role is responsible for recruiting talent to the team. In this role, you understand the importance of bringing in new hires that are both well equipped to do the job and a strong culture fit. This role manages all efforts to recruit and interview top talent. Over time, talent acquisition can be broken into even more specific roles, such as department-specific recruiters or recruiting operational positions.
- **HR data analyst:** this role tracks metrics throughout the talent lifecycle to inform people of decisions about employee programs or processes. This role is still relatively new in the field, so it requires a lot of creativity and strategic data-driven thinking in order to develop useful and actionable metrics.
- **HR/benefits administrator or manager:** employees' primary communication with their people operations or HR team typically centers around compensation and benefits. Whether it's a missed paycheck, open enrollment, or annual reviews, employees rely on people operations or HR to answer their questions and make sure everything is in order. This role maintains processes to ensure that employees are properly rewarded for their work.
- **Employee relations manager:** employees need a central contact for their disputes, professional concerns, and questions. Within the people operations/HR department, the employee relations manager handles all these matters. They work collaboratively, acting as the

point person between the business and people OPS/HR team, relying on peers for expert guidance when different employee situations arise—but their primary focus is to advise and consult employees/managers on any employment concerns.

HR Skills Needed

- Human resources, organizational development/behavior degree if you want to jumpstart into leadership. Otherwise, a degree is not needed.
- Active listening, verbal/written communication, organizational and analytical skills, detail-oriented
- More importantly, humility and eagerness to learn and serve

Sample Recruiting (Talent) Coordinator Career Description

Coordinating roles is one of the best ways to break into the people and recruiting side of a business. As you'll see in the sample career description below for the talent/recruiting role is definitely one of the least technical gateways; few priors skills are needed.

A talent/recruiting coordinator's central role typically is to automate the manual process of creating interview schedules that work for both candidates and interviewers, and to ensure the next steps and feedback from interviews are coordinated properly. I've made bold, specific points in the following career description to draw your attention to the most relevant aspects of it.

What You Will Do

- Ensure a world-class **candidate experience.**
- **Coordinate the recruitment process** and get experience interviewing, evaluating skill level, **taking and giving feedback,** and onboarding.

HUMAN RESOURCES (HR) IS A STRATEGIC PARTNER TO LEVERAGE YOUR PEOPLE TO HAVE FULL ADVANTAGE.

— PAMELA MCEWIN
TECH STARTUP HR MANAGER

» **Manage the summer internship program** for our New York and London offices.
» Establish yourself as trusted partner and point of contact for hiring managers across the organization.
» Create captivating job descriptions to help **articulate and market** each unique career opportunity.
» Build relationships with relevant industry professional associations, universities, meetups, and other appropriate events.

What You Will Need

» The foremost quality for this or any position is **integrity.**
» You must be self-motivated and operate with a **high sense of urgency** while working effectively within a team environment.
» You must have a proven track record in the following abilities: organization, motivation, and **follow-through.**
» You're a **charismatic people person** who can talk to anyone; you're flexible, fearless, and excited to help build something awesome and share it with the world.

Two Journeys into Tech via the HR Gateway

We're going to learn about two individuals who both entered tech through the HR gateway. The first is Marisa, who started in retail, left college early without graduating, and worked in multiple industries before breaking into recruiting. She started as a recruiting coordinator at Cloudera and jumped to doing recruiting at Facebook and now Slack.

The second individual is Aisha Stephenson who took a different approach. Aisha studied American studies at Stanford before getting her masters in human resources. She filled three different HR roles before becoming an executive.

Marisa's HR Journey: Recruiting at Facebook

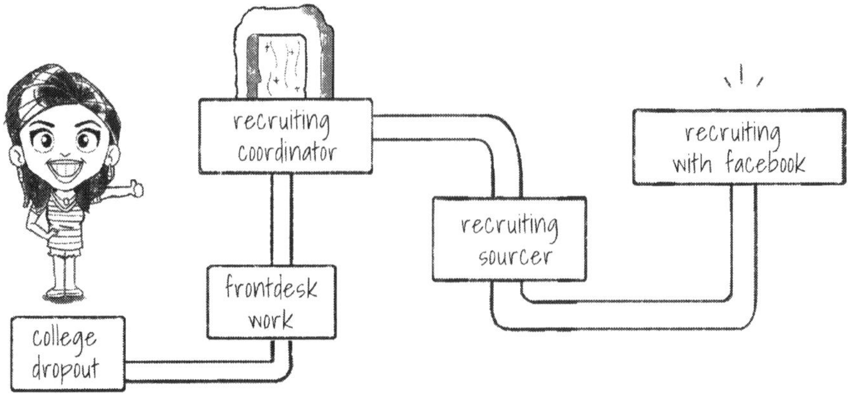

Marisa decided to marry her college sweetheart and move across the world to San Francisco. As if this wasn't scary enough, in order to move, she had to drop out of college. Oh, and she was only 21 years old.

After a few months of living the housewife life, she wanted to prove not only to her family and friends but to herself that she could be successful without a college degree. Marisa took the first step and began to build her resume. She quickly became discouraged. Questions of doubt started to fill her mind. "Should I say on my resume that I never finished school? Will anyone want to interview me after reading that?"

Marisa reached out to a local staffing agency to help her navigate the San Francisco job market and to give her guidance on how to best approach her lack of college degree. Her recruiter, Amanda, was extremely helpful and focused on getting her jobs for lightly experienced candidates. Amanda pitched her a couple of roles she was looking at. Boom! Marisa got three interviews that next week.

One of these jobs was at a chiropractic wellness center as an office manager. Marisa decided that this was the best culture fit and gave her the best visibility to the other job functions in the office to see what she did and didn't like. In this role, Marisa learned the backbone of business operations for a small company and how to manage client relationships at a micro level. After a year and a half she left to pursue other opportunities.

Marisa quickly noticed that this time her job search was a lot easier because she'd gained experience with transferable skills.

Her next job was working for Stitch Fix as a virtual stylist. She enjoyed the "work from home" freedom, but this still wasn't what she was really looking for as she wasn't growing as much as she wanted in the role.

So, Marisa decided to reach back out to her original recruiter, Amanda, and pick her brain about the recruiting industry. Amanda referred her to someone who was looking to fill a role with a finance firm to assist with HR, recruiting, and office management for their small 10-person company.

By this time, Marisa was good at selling her experience, so she nailed her interview! But after a few weeks there, she realized she was more doing responsibilities of an executive assistant, coordinating corporate parties and booking flights for candidates, washing dishes, etc. And that was when she had an aha-moment: the smaller the company, the more likely they are going to have less segmented roles. So, she decided to keep on swimming to a larger pool where she could be more specialized and could transition out of the office management role.

Because of her experience thus far, Marisa knew she wanted to pursue recruiting in a larger sized company. So, she compiled a list of her ideal companies to work for: Facebook, Airbnb, Pinterest, and Cloudera. Within these companies, the specific role she applied for was called a recruiting coordinator. Their job junction was to support between 3 and 5 sourcers and recruiters, and after they got the candidate interested in a role, they would help schedule all aspects of the candidate's interview and serve as their point of contact when during their onsite interviews.

Long story short, she crushed the interview with Cloudera and started the next week.

Marisa started her journey in recruiting: she learned how the recruiting machine worked. The sourcers had a number of candidate conversations they had to produce every week. The recruiters also were expected to build a relationship with the hiring managers, so they really understood their needs and wants in their ideal candidates. This allowed them to pre-sell the candidate and explain why the hiring manager should take productive time out of their schedule to speak with particular candidates.

Marisa was super focused on wanting to crush it as she finally found a job that didn't involve cleaning or managing frustrated customers. She got promoted to a sourcer in 6 months after starting with Cloudera.

This is how her journey as a talent warrior began. Every week their numbers were shared on a Google sheet, and she had to confront her deepest fear of failing. She started to build relationships with her recruiters and hiring managers, and made sure she learned from them to understand what the business and team needs were.

After the first quarter, she moved up the pack, and by the second and third quarters she was the #1 sourcer in the company! Marisa had a different fear now, which was how to stay #1. Eventually, the game became less challenging. She found it was easier to find and book candidates quickly, and she was yearning for growth.

Internal Champions

One of the main takeaways from Marisa's journey that I want to point out is the concept of internal champions. This is why Marisa was promoted so quickly: she gained the respect of her team, client partners, and managers. In doing so, she built a team of internal champions. These were the folks who would fight for her when she wasn't in the room and who had some sort of influence on the executives who were in charge of promotion approvals.

She did a great job building multiple layers of champions: her own recruiters, the manager of sourcing, the manager of recruiting, and client partners (the people who hire her candidates). One of those internal champions was able to fight for her even after he left Cloudera for Facebook when Facebook was reaching out to Marisa for a sourcer opening.

In the interview process, Marisa did another amazing job impressing the hiring board. When it came to final decisions to have an offer extended, they checked her references, including those of some of her internal champions. That's what helped seal the deal and landed Marisa a job at Facebook.

This is the power of starting as a recruiting coordinator if you want to break into recruiting. The job is simply scheduling candidates with the hiring managers and other interviewers for their interviews. With the right focus and hustle you can transition into sourcing and build your skills and internal champions to become a recruiter.

This can also lead you to leadership opportunities within recruiting or even venturing on your own and doing your own headhunting for other companies, like the staffing company who helped her get her first job in San Francisco!

Marisa's story is inspiring! She isn't what you think of when it comes to folks in technology. But what she lacked in credentials she made up with drive and coachability. If you ask her hiring managers, they'll attest to her keen understanding of the client (hiring manager) and the talent side of the recruiting business such that she delivers results!

GOLDEN NUGGET

To return to the importance of building internal champions: if you ever wondered why you were not promoted or why someone else was, it boils down to not knowing who the real decision-makers are and not having enough or the right internal champions to fight for you. In my world, internal champions are those in an organization that have both influence (aka power) and are fighting for you and your cause when you are NOT in the room. Building champions is definitely something to take away from Marisa's story as she built them at every level—employer, recruiter, hiring manager, peer levels—which helped her get into Facebook.

Aisha's Take on the People Team in Tech

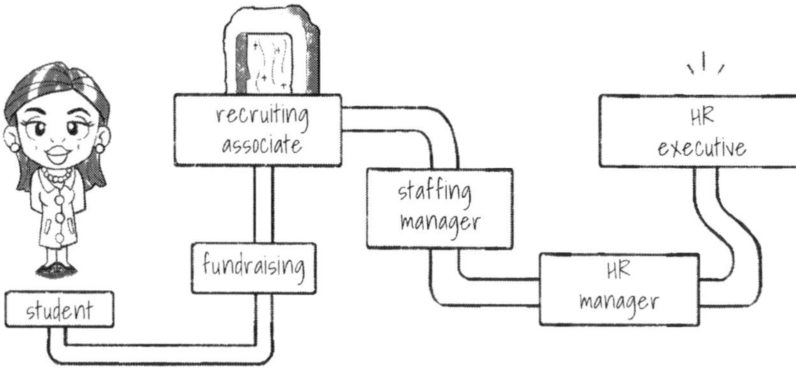

I want to add a perspective from an executive here as the people team is much more than just recruiting. Aisha and I both started at the same time at OKTA, and she led many people organizations in her career. At the time of the interview Aisha was leading the people team at a startup called Quizlet. Her interview will give a much clearer idea as to what it takes to succeed in the people business and the ideal type of candidate.

Aisha: When I came out of college, not having any idea what I wanted to do, living in the Bay Area that brought me into the tech space. After college, I saw all my friends going off to tech companies or consulting firms. My first job after graduation was at the Stanford Fund and my role focused on fundraising for young alumni, and I hated it. Working with alumni in support of scholarships for students was fulfilling but I hated asking people for money.

So, I got great advice from my friend's father who's a retired headhunter [executive recruiter]. He told me, "Don't focus so much on the titles; just focus on the duties and responsibilities of certain roles. Look at it that way because I guarantee you will find something in the duties and responsibilities that you like to do or that you think you'd be good at." And that's what I did.

I actually looked on Craigslist, back when people used to use that, and I found a role for a recruiting associate at a company called PlanetRx, which was the competitor of Drugstore.com. They took me on their team, and within six months I moved from recruiting specifically to HR in general. I was promoted to manager of staffing. In this role I worked with everybody—from front desk to the VPs to our customer support team—and that was the most exciting thing for me because while I didn't see a future for myself in sales, marketing, or engineering, I could see a role in HR that allowed me to work with every single person across the company from bottom-up and top-down. I decided, "I really, really like this!"

Irwin: What do you think is an important quality to have when working in HR?

Aisha: It's really important that you are a people person, and by that, I mean personable, warm, and inquisitive demeanor. You want to get to know a person as they move from candidate to employee. In any organization that I work for, if recruiting isn't reporting under me, along with HR, then I absolutely make sure that there's a great rapport between the recruiter and me.

When we sign on new hires, they come in thinking that it will be a great place to work. We want to make them feel even more so as an employee, as a team member. From their first day you want to make sure they get the lay of the land and that they're excited. That typically happens when there's a great handoff from the recruiter to the HR so that you can maintain that excitement that they had in the interview process. Who's going to guide them through their first day and get them oriented? What's the ramp plan? What's the onboarding plan? What are the get-to-know-you opportunities that are coming up? There has to be a seamless transition from candidate to new hire, or else you could lose the opportunity to retain the employee beyond the first year (if not six months).

Irwin: What is the most interesting part of starting in HR to you?

Aisha: In HR, people operations, or recruiting, you can effectively come into an organization and have a career ladder or path as it relates to human resources. In the startup or tech space where everything moves at an accelerated pace, you can start out as a coordinator and go to a manager in half the time that you could if working for a bigger company.

People Operations - it's really about being driven and passionate about people while being able to thrive in dynamic, and fast paced environment. When you start, you show that you are willing to do whatever it takes, you are willing to figure out something that you don't know. You can crowdsource information and figure out best practices and then apply that to building your people programs. It doesn't matter if you go to college or not because you can enter the people operations function and use your experience to work your way up the career ladder over time. I would say it's helpful to be analytical and business-oriented if you decide to go down this path.

Irwin: Where would you start when breaking into HR?

GOLDEN NUGGET

> **Aisha:** Research the company. It's really important to know who they are, when they started, what the main business focus is, so you can discuss that in depth during your interview. You'll be able to show how you would like to be part of the organization because of its mission, industry, product, or services. You can also get a great understanding of the whether or not the company is a 'people first' organization by reviewing their website and how the job descriptions are written.

Be ready to demonstrate how you take initiative and show that you've done your homework on the position and company. You can start out as an office manager or a recruiting coordinator. It's an entry level position, but it's a way to get your foot in the door, so you can show that you can be a key contributor: "I can be a team player. I can really add value." That's what they're looking for, beyond experience and expertise.

And never underestimate a sunny personality and disposition, one that can relate to all types of people and get people excited, a real champion for the company—that helps.

Irwin: Would you mind sharing the different career progressions and give advice on how a person can best move up in an organization? How long would that take? And how could the speed of a company growing matter?

Aisha: It's interesting because when I first came to Quizlet two years ago, the recruiting manager talked to me about a possible transition to human resources, specifically moving away from recruiting. I said that if you want a career in HR, there are a lot of different ways you can go about it. You don't technically have to have a lot of experience. Go to a bigger startup because there are more opportunities for you to move from the talent acquisition to talent management team at a similar level than from coming in as a HR assistant or coordinator, or a recruiting coordinator.

There's also HRIS tech (Human Resources Information Systems). I have somebody that I hired at OKTA who came in as an HRIS associate. That role entailed effectively taking all the employee data and inputting it into our HRIS.

There are more ways if you have a bigger startup or company. At Quizlet, a smaller startup, the people operations manager joined the team two years ago as the office manager. That was two years ago, and she eventually got promoted HR manager by leveraging her Quizlet institutional knowledge, getting involved in HR projects, and taking classes on HR fundamentals.

Also you can get certified in HR.

Irwin: What is the certification?

Aisha: It's called SHRM, which is "Society of Human Resources Management." I actually got my PHR a couple years after I started. It's a way for you to get that foundational knowledge while also working.

This person who became my HR manager went to class one night a week for three or four months. She is now ready to take the test, which gives a certification either as a professional in human resources or a senior professional in human resources. Effectively, it's like taking all the knowledge and experience she gained as an office manager, coupling that with the foundational knowledge she got and the functional expertise. Now she's actually getting involved in employer relations issues.

Another way to break in is through talent acquisition or recruiting. That path is usually coordinator to junior recruiter to recruiter—either on the technical side or on the business side—senior recruiter, recruiting manager or principal recruiter (individual contributor), and then director of talent or talent acquisition or recruiting.

You can go the generalist route, which is what I took. You get steeped in everything. When you come in as the first people operations or HR and recruiting leader or point person, you can effectively manage both of those sides, so you'll have, generally speaking, a recruiter that you hire and then somebody who can work with you in terms of HR administration benefits that you hire. Getting them set up with equipment, workstations. Those are three different ladders in and of themselves.

At a bigger startup or bigger company that has been established for a long time, you can come in strictly through benefits. You can come in strictly through compensation. You can come in as a compensation analyst. I know a lot of people who were consultants. They actually switched to HR people operations, effectively coming in on the analyst side.

Irwin: You mentioned certifications. To stand out from the crowd, would it help to have a certification ahead of time?

Aisha: Yes, it's helpful. I fell into HR; I was probably three or four years into HR before I went to get my certification, went through training, and then took the exam and got certification. I know a lot of people that take the training first and then effectively add that to their resume, which is really, really helpful.

On most HR job postings, they actually ask for a certification in HR, PHR, or SPHR as helpful, preferred, or A+, but you don't need it.

Irwin: Do you need a college degree?

Aisha: No. Don't get me wrong; I really enjoyed my college experience, but we've seen examples of successful people in the tech space who didn't go or get their degree. Mark Zuckerberg's a testament to that. If I recall correctly, he dropped out of Harvard after his sophomore year to start Facebook.

Verbal and written communication skills, especially in HR and recruiting, are so important. You're writing job descriptions or writing communications about rolling out a new people program or process. You do need to have experience and practice writing to large groups of people. So, college is helpful, but not necessary. Just like the certification: it's helpful, but not necessary. What you really need is the passion and desire for it.

Irwin: How would you advise someone who's maybe not feeling really passionate?

Aisha: If you're not passionate about it, you shouldn't do it. You know HR is often a thankless job. There are many times when you can feel a little discouraged that the general team, the company isn't regularly acknowledging the efforts that your team is making on a regular basis. There's a lot of work that goes into recruiting and retaining employees, and many people don't see a lot of what happens behind the scenes. And HR is typically the complaint center. So, if you're not passionate, if you don't really, really like it, then move to the accounting side of things, payroll, recruiting. However, if you're specifically going to be on the HR operational side, you do need to have a passion for it or at least a commitment to stick it out because there will be those times when you have conflict come up, and it's hard to push forward without that passion.

You cannot take sides. Some people will say, "HR is always the advocate for the employer." I think HR is the advocate for the employee AND the employer, and that's a really, really thin road to walk sometimes. It's challenging, so you have to have passion for it.

Irwin: Are there any questions that I have not asked that you think I should be asking you?

Aisha: Let me add that HR has been good to me, although I have had to dig deep sometimes to keep coming in daily with a smile on my face ready to face another day. Supporting people interests me. Having people learn something new on the job about how to collaborate better, about how to be more productive, about how to be a first-time manager, that's all fun for me to create, coordinate, and organize. Those times when people don't like the snacks in the kitchen or they don't like that we don't get that holiday off, and they come and complain to me as though I did it—those are the times when I dig deep and reflect on more positive situations when people were really, really excited about reaching a company milestone and we had a celebration; or somebody was so excited that somebody recognized them on their work anniversary and recognizing them for all the hard work and saying, "Thank you so much for everything that you do." Those times are what get me through the more challenging times, and they are why I'm still in HR and excited about the future of the People function.

*

If you have a passion for people and seeing people come together and supporting them, then HR and the people team is your family! I have seen many folks from non-traditional backgrounds grow in these people-related gateways. The people team sometimes laments that they have a thankless job, but when you have the right team, they do make everyone's work-life fun and connected. I hope you are seeing that each gateway into tech offers roles to different personality types, and some career paths can cross into others! Therefore, if you are interested in a tech career, but you have no tech experience or degrees, it doesn't matter. There are still plenty of important and interesting jobs for you in the tech world.

In the next chapter we are going to look at another critical and often overlooked gateway: the customer support gateway.

CHAPTER 10

THE CUSTOMER SUPPORT GATEWAY

The Shep Hyken quote that introduces this chapter, "Well-executed customer service encourages customers for life," is a simplified explanation of how important taking care of the customer is. Businesses invest a tremendous amount of effort, time, and money in acquiring a brand-new customer. One of my mentors asked me, "What would happen if all your customers who bought from you in the past never left? What kind of impact would that mean for your business?" His goal was to help me understand that in order for a business to flourish, repeat and happy customers are key. Customers who come back again and again are important and help create the foundation from which the base is built. This is why subscription-based model are so popular in tech companies today. There is a need to increase how much the customer spends over their lifetime (CLTV: customer lifetime value).

A customer's success before, during, and after they buy has become one of the top corporate values in companies these days. This has led to a tremendous opportunity for folks who want to break into technology. For example, in technology B2B business there is pre-sales support, where is help before the buyer becomes a customer; there is also help during the service after the customer has purchased it, known as customer support; and there is a team to help deploy and remedy troubled customers' issues that have escalated to a point where it requires a specialized team called the customer success team.

For those who are in or who enjoy service-related roles—problem solving, serving, helping, teaching, etc. —please pay attention because this gateway is for you!

The customer support career segment is always growing, needing, and churning because people are leaving or getting promoted out of their roles. With the growth of technology, this gateways allows one of the simplest ways to break into technology and one that also offers a career path to other amazing roles.

Gateways

- Customer support representative/specialist who works with customers during sales ($39k–$62k)
- Customer success representative/specialist who works with customers after sales ($54k–$116k)

Career Paths:

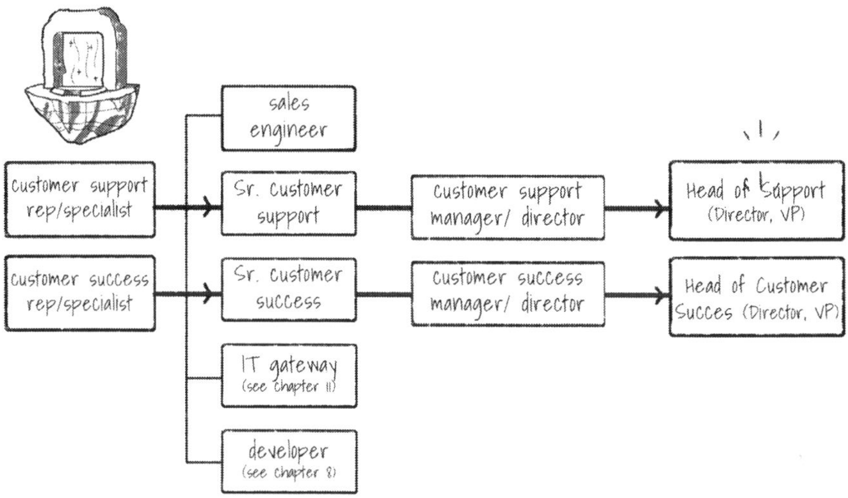

Here's my breakdown of the customer support versus the customer success roles:

» **Customer support:** this team is more reactive in nature as it's the team that responds to and resolves a customer's problem that are being issued by the customer themselves.

» **Customer success:** this team is proactive in their approach. They build relationships, understand their initiatives, and work towards helping the customer reach their goals.

Customer Support and Service Skills Needed

- Listening to the customer's need
- Problem-solving skills
- Ability to prioritize tasks and manage time
- Patience and having a servile attitude

Sample Customer Support Agent Description

The job description you'll find below from a large crypto exchange's career page is a two-birds-one-stone type of opportunity. First, joining an exchange would automatically allow you ride the cryptocurrency wave (see bonus chapter: Breaking into Crypto) and second it would allow you to ride a wave of massive hiring. At the time of writing the CEO just announced that they are hiring for more than 200+ additional support roles due to the overwhelming amount of support tickets and new user registrations that are rolling in! It's worth noting that this gateway requires a balance of work hard and the ability to empathize.

Notice when you read the description below, the position entails answering tickets on weekends and being very patient with customers. This gateway is one of the most in-demand roles that doesn't require a technical background. ost people who start here answer basiclevel 1 type of support questions. You can learn on the job and eventually pick up more skills to help solve more difficult problems. This gateway can very demanding but very rewarding as many of my friends broke in this gateway and had a great career moving up and laterally within an organization. The best candidate for this gateway is one who is willing to serve, and problem solve. If you are used to working in a service related role, this gateway is a very good one to transition into tech.

At the very end of the career description, notice it says that knowing a foreign language is helpful. Take a look and see if this type of gateway fits your personality.

Responsibilities

» **Manage, prioritize,** and **respond to inbound emails**

» **Identify common trends** within support emails to preempt issues and bugs

» Monitor transactions occurring on the platform, **identify fraudulent activity,** and **correct transfers** when needed

» **Collaborate** with the product engineering team to **identify which issues and bugs** require immediate attention

» Exercise sound judgment when corresponding with customers

» Utilize email to communicate with customers

Requirements

» Great with people—customer service/technical support experience a must

» **Ability to empathize with and educate** customers about digital currencies

» **Self-motivated**—ability to work independently and remotely

» Excellent organizational and written communication skills

» Experience with supporting financial services or digital currency

» **Advocate for digital currency** and **passionate** about building the digital currency community

» Craving of the opportunity to master a subject and teach what you know

» Strong **attention to detail**

» Experience dealing with **time-pressured situations** and high-value accounts

WELL-EXECUTED CUSTOMER SERVICE ENCOURAGES CUSTOMERS FOR LIFE.

— SHEP HYKEN

- **Problem solver**—you don't mind the occasional "rabbit hole"
- **A sense of humor** and a strong personality
- A sixth sense for avoiding social engineering attempts
- **Willingness to work at least one weekend day per week**
- BONUS: **fluency in a foreign language other than English**

Jim's Take on Customer Support and Success

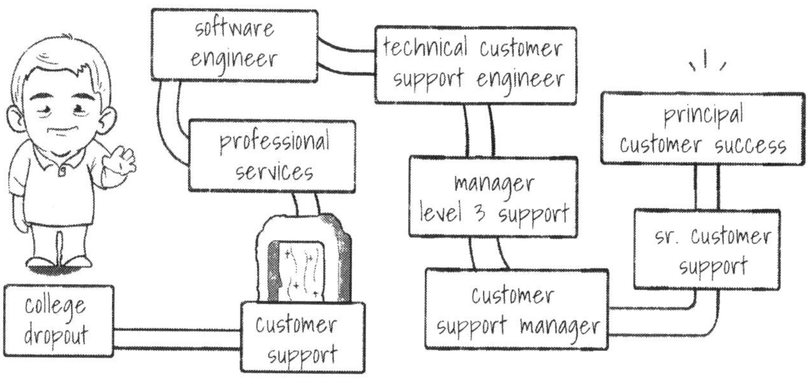

In the interview below you'll read of Jim's journey into customer support and customer success. Jim and I worked together at OKTA in the early years. He added so much value to the customer support side, which in my opinion was one of the toughest jobs. There were hundreds and hundreds of supported tickets that he had to cater to, and I would see him at work early and continuing to work late to take care of our customers. The best part—he's another person who broke into tech who had NO college degree. His story will inspire you.

Jim has moved on from customer support into the customer success side of the business in the Southeast USA. When I was thinking as to who would be amazing at sharing their journey into technology through support and customer success, Jim was definitely my guy.

Jim: I got my foot in the door with any kind of a job that I could get with a tech company and worked my ass off. My grandfather used to tell me, "I don't care if you're sweeping the floor at a place or if you're running the place, always do your very best, and you can't go wrong."

So I did that, I got in there and moved throughout that company. I started running around and fixing people's computers and eventually moved into managing that part of the company. Because of the work, I did there, I eventually became a software engineer/developer, which seemed like a good progression to me, but it really led me into a place that I wasn't getting as much satisfaction out of the job as I was when I was actually out there helping people.

Always making a move is not always the best thing. I still did well at my job, but I just didn't really like it anymore. One of the benefits of that role were the relationships I built with the people there. One of those relationships I built was with one of the software developers that came to work for OKTA early on. His name was Scott.

Scott basically called me up one day and said, "Hey, I've got a job with your name on it. Why don't you come in and apply?" So I did. And sure enough, it was a great fit. I started off doing the customer support. That was in their early days, so we were extremely busy, extremely short-handed, and I worked long hours.

Irwin: You started as a support specialist from the very beginning at Cable Data. You did that for a couple years. Then you moved into the IT side first. How did you land that job initially? Why'd you pick that role?

GOLDEN NUGGET

Here's a great example that resume's can get lost. What I learned from Jim is the importance of getting in front of the hiring manager.

Jim: I applied for the job because it seemed like a really good fit for me. After a few days, I didn't hear back from the manager. So I drove down to see him in person and said, "Hey, I applied for that job, and I haven't heard back from you. What are my next steps?"

Irwin: No way.

Jim: Yes. I brought my resume with me and walked in. He looked through it, and he said, "Hold on a second." And he called HR and asked, "Why haven't I seen Jim Knutson's resume?" And he continued, "Well, he's standing here, he handed me his resume, he looks perfect for the job. So I'm sending him over. I need you to sign him up." You have to be persistent, but not to the point where you're gonna get a restraining order.

Irwin: How did you even know where to find the hiring manager, first of all?

Jim: I knew the company, and I knew he was running IT down there. It was a new department that they had just started. This was 1989, so we didn't have email; we didn't have cell phones. I couldn't look him up on LinkedIn.

I just went through the company directory, found out who he was, what he was doing, and said, "Why haven't you hired me yet? I'd be perfect for this." Which was true. It took some initiative to do that, and I was very, very young at the time, in my twenties. So it was just one of those things where you just felt like this is right, you gotta do it, and you just go for it.

Irwin: You worked your butt off. Then you moved into supporting the IT side. What was the next progression after that?

Jim: After that, I went into professional services to support our customers at Cable Data. Professional services was a great job. It was kind of like supporting our internal folks with the exception that they're paying us to do it. Tons of travel, so that was good. And just helping them out with solving their problems. It could be a customer who was anywhere in the United States pretty much. I hit every single state with that job. And it was great. From that I decided I wanted to go into software development for some reason.

Irwin: At DSC.

Jim: Yes, DSC is the same company. That was a 20-year career span there, working for the same company.

Irwin: So you learned how to code? How'd you learn how to be a software engineer?

Jim: I went to the Institute of Technology.

Irwin: Oh, you didn't go to college?

Jim: No, I didn't go to college.

Irwin: No way. You're the fourth person I've met who didn't go to college and got into tech. You did IT support, then you moved to customer support, and then you went to software engineering. How long were you doing software engineering?

Jim: Eight years or so.

Irwin: Oh wow, you did it for quite a bit. How would you describe the differences between customer support and customer success?

Jim: In customer support, you're putting out fires. Customer success, you're making sure those fires never come up in the first place, often by engaging with your customer regularly, even when they don't need something. It's almost like a sales role, except I don't sell anything. It's different than support because support's always reactive, whereas success is a bit more proactive.

Irwin: Let's talk through your OKTA story. So your friend referred you in, you were the perfect candidate. Why did you feel that you were the perfect candidate?

Jim: Because of the job description that they put up that included things like "able to handle multiple projects at the same time," "able to work and interface with customers," and "able to solve complex problems." It just looked really cool.

Irwin: I don't want to glamorize what you do, but you love what you do.

Can you walk through it, so we don't glamorize it, and tell us what makes a great support personnel, what you love about the job, and how you go through the job? Because it's not one of those easy jobs, but it's one of those definitely needed roles.

Jim: It was pretty demanding for most people. We did burn through a couple support engineers. It's high volume. People are calling in, they're upset, angry, frustrated, don't know where to find their answer, and you could spend five minutes with somebody finding their problem and solving it for them. At the end of the call you've done two things: you've changed their attitude, and you've changed their day. And you've also educated them a little bit because they're not going to call back about the same thing twice, usually.

Irwin: Okay, so you just like helping people?

Jim: Yes, I like helping people.

Irwin: Could you walk us through a day in the life of being in customer success, what life is like there, and the different skill sets that you need to be successful in that role?

Jim: No two days are the same in customer success. You might be putting out a fire like you do when you're in support. You might be educating somebody, showing them some new features on a product. You might be going through a usage review on how they can get the most value out of the product, what features they are entitled to that they're not using. There's so much to customer success, making sure that your customers that are using the product are staying successful with it. Also, there's always the chance to pitch customer success to another customer. And so we've got that as well. It's multifaceted: the fire-fighting aspect, the account maintenance aspect.

Irwin: I hear there are three different aspects to what you do. One is support, supporting a customer that may be dealing with some issues. Second is proactive support where you're proactively making sure that they're using the product to the fullest to ensure that they renew, for the most part, or are happy and continue to be a customer. And then there's a third aspect, which is there's a new customer and maybe it's a large customer, and you need to educate them on the value of buying some sort of implementation or high-end support so that they are successful in their trail. Is that a good summary of the three different things that you do on a day-to-day basis?

Jim: There's a little bit more to it, but that's a good summary.

Irwin: In your opinion and in being in this side of the business, what are the different ways people have broken into technology?

Jim: Customer success is one of the major entry points. It's surprising to me because it doesn't really seem like an entry point. But we've brought people from customer success over from professional services. They have a PS (professional services) background, they're used to dealing with customers, that kind of stuff. We've even brought people over from sales into customer success. And then obviously from support and those kind of backgrounds for customer success.

Irwin: How do you think people get into professional services then?

Jim: My gateway for professional services in another career was through desktop support (see chapter 11 The IT Gateway). I went from desktop support to professional services. You get into customer support, and that's a gateway. And that's entry level, anybody can do that if you know the product and just how hard you are on the product. And then obviously if you're good at what you do, you get a chance to advance into other opportunities.

Irwin: Any advice on the skills? If a person was to go from zero tech background to customer support, what are the things that they can learn ahead of time to stand out, and what advice would you give someone looking to break into customer support and stay, or have the longevity to stay through a customer support?

Jim: When I was building up the customer support team, one of the things I would always ask my candidates to do would be to download a demo version of the product and install it. And so I would learn A, their skill set, and B, whether they follow through a task when they're assigned. They didn't have to get it perfect, but as long as they made an attempt at it, and then you talked about where they ran into trouble, that kind of stuff. And that told me their ability to troubleshoot it. It was a really organic way to interview people and get a true sense of what their skill set was and how good they were at it.

Irwin: It was a very active interview. Go download the product and play with it. Any advice for skills to learn for future trends? Any trends that you see that'd be interesting in the tech space in general?

Jim: Focus on some core things, such as listening to your customer. What are they actually telling you? Read their emails, which is huge.

Irwin: Do your job.

Jim: Yes. Some of these guys with support like to close tickets and that's what they do, I get it. But you have to give time to listen to your customer. Listen to the customer, determine what they are really saying. Listen and give back your guidance to your customers when you're working out in the field and know something. There's never a bad time to say, "I don't know the answer to that, but I'm going to go find out. Then I'll let you know," even if you think that it's something that you should know. If you give the wrong answer, it does not build confidence in you as a support person. And it does not build confidence in your relationship with that customer.

Irwin: What are your thoughts on how do you break in a tech if you're not in the Bay Area?

Jim: You've got to build a relationship and build some trust with an employer, and then you should be able to do the job from anywhere, as far as customer support goes. You've got metrics, so we can measure to make sure everything is going the way that we want it to. There's no reason to come into an office. And I can see that being a trend in the future as well.

*

Jim is a wealth of knowledge and inspiration. His story of knowing what he wanted and persistently reaching out and finding the decision-maker for this breakout career is a key takeaway. If you like helping other people, you can build a career starting at the customer support and desktop support gateways (see chapter 11 The IT Gateway). These gateways are in high demand and are needed at every company and lead to careers and places in the world you may have never dreamed of. It can be demanding, but it's a great way to learn how to problem solve and prioritize your schedule. Either way, you'll definitely learn something.

Insider Secret: Sales Engineering Career Path from Support

Have you ever wondered how account executives talk to very technical buyers or answer deep technical questions? Have you ever wanted to be in on the action of a sales deal without having to be on the hook for a sales quota?

I want to introduce you to one of the coolest jobs in the business: the sales engineer (SE).

Most companies that sell to technical buyers have sales teams that are composed of two parts: an account executive (AE) and a sales engineer. The AE, as discussed in chapter 3 The Sales Gateway, is responsible for managing and building new relationships and ensuring that the right resources are pulled into the deal as needed. The sales engineer is the account executive's technical counterpart. The SE's responsibilities include answering very technical questions, giving deeper technical demos, and if the customer wants to test the solution, the SE helps ensure proof of concept (POC) in order to resolve any doubts around the solution not working as designed. SEs are in on the meetings and play a huge part in ensuring a deal gets made. AEs can't survive without them. The AE focuses on managing multiple deals, and the SE focuses on troubleshooting and answering technical questions. They help automate a lot of the technical gaps an AE will have.

I think the SE position is one of the coolest jobs in tech. They are paid really well ($90k-$166k not including commission) and are in on the action of working with a customer during the sales process. They don't carry the weight of a quota but have a component of their compensation tied to the overall team number!

Only Getting Bigger

The customer support and success gateway is powerful for breaking into some of the best tech companies in the world. Brian Armstrong, the CEO of Coinbase, has stated that he has scaled his customer support team from 40 agents to 220 agents, and by the end of the year, there will be an additional 400 agents on top of that. The takeaway for you is that customer support is a role that non-techies can thrive at.

Support is by far one of the most forgiving ways to break into technology, as it requires the least initial technical experience. You just need to problem solve, learn on the job, and grind, as Jim mentioned. However, it is one of the more difficult roles operationally as it involves constant firefighting. If you look at the long term, you can see that this gateway can lead to many different roles within an organization.

Jim hinted at the next gateway. When asked how he saw people break into customer success, he noted that a lot of folks start their career in desktop support, meaning IT. The IT people are the ones that I, as a salesperson in tech, have been selling to my entire career. They are the folks who are responsible for managing the IT infrastructure so that the business can run smoothly. Think the corporate phone, desk, email, software applications, security, etc.

CHAPTER 11

THE IT GATEWAY

Have you ever had to deal with a slow computer, spotty internet, or an uncooperative printer demanding an ink cartridge change ASAP? Yup, this technology has become a key part of our lives and with it comes changes, challenges, and maintenance. Lucky for you, as a company grows they need to hire an IT (information technology) team that handles all these issues so that you and everyone else can do your jobs.

This is one of the most in-demand careers, yet one of the most overlooked fields, which means this is your opportunity. Think about when you start a new job and part of your orientation involves someone on the IT team getting you a laptop, ID, phone, email, and wi-fi. This is only what YOU see.

What isn't seen is what it takes to run a network for the entire global 100+ locations. Or creating data centers to run the internal applications that the company is developing to stay competitive. Or the security initiative to ensure that the proper procedures and security tools are in place to ensure the finance team doesn't wire money incorrectly and that prevents 20,000 employees from clicking on the wrong link that could release a virus that could lock up your computer and the files you are connected to (google: ransomware).

In order for a business to reach its full potential, the IT team must be strong and efficient. Here's the best part: you don't need to have prior experience or all the skills I previously mentioned. All you need is the desire to learn and work hard. Oh, one more thing: you must have a passion to learn how technology works together. The IT gateway is similar to the previous chapter about the customer support gateway. The main difference is instead of supporting customers, you'll be serving your internal employees. This is the way many people in IT get started. It's the gateway I've met the most number of folks without a college degree in tech. Let's dive in.

Career Paths:

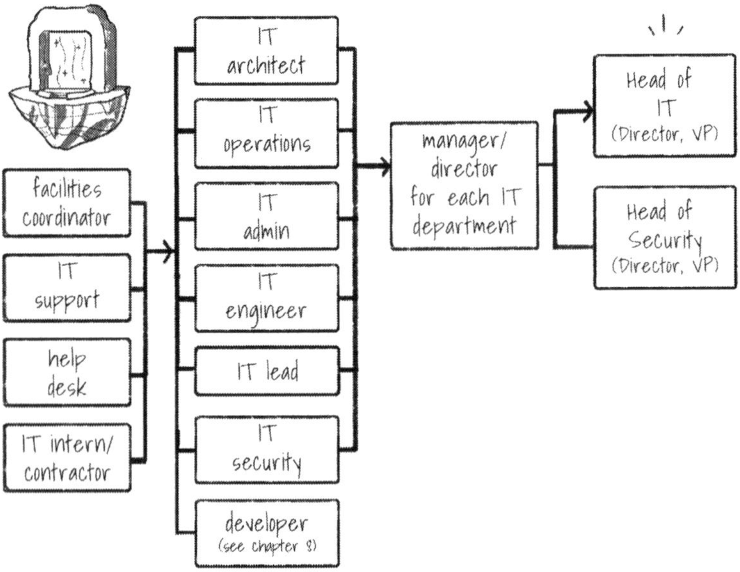

Gateways

- Facilities, IT support, helpdesk/desktop support ($33k–$65k)
- Customer support representative/specialist ($39k–$62k)
- IT contractor/intern (n/a)

IT Simplified

- Internal IT (operations): the IT team that helps make sure you can do everything when you go in the office. They set up your email and wi-fi. They are the people in the office you go to for computer help.
- External IT (platform): the IT ninjas who support your external facing apps. For example, if you buy stuff on Amazon or at Sephora, there is a separate team focused on ensuring the IT infrastructure behind that is reliable and up.
- IT security: the person who ensures you can't get the company in trouble when you decide to open up malwareXYZ.com or decide to click ALL the links and attachments in your email

Skills Needed

- Passion for technology
- Willingness to learn
- Work ethic

Sample IT Helpdesk Administrator Career Description

Yup! That's the skills needed to start folks. The IT department is a service-oriented department. IT is there to serve the business to ensure it can run at full speed. As you'll see from the career description below for an IT position. To land an IT position, you must be very interested in technology as almost all of it is learned on the job. You need to be a quick study, show that you came to play ball, and support the business. Most of the tech companies I research wanted 1–2 years of prior IT experience. I would try to get that experience any manner possible and then pursue a job at your tech company. Things I like in this posting are the clear 3-month and 12-month visions of what the person who takes on this role should achieve. I've bolded a few key points to pay attention to, such as how the phrases, "Being fluent with, learn applicable systems," shows how a key skill is to be able to explore, play with, and learn new systems as part of the job.

What You'll Achieve

At 3 months

- **Support the End User** community for all real time requests
- **Learn applicable systems** with the help of the team so that you can be a master at the most common support requests with escalation support from our Regional Leads and Operations Engineers
- Use your tele-presence expertise to assist our IT team with LifeSize/Zoom video conferencing across the global company and maintain the production video technology for internal All Hands sessions

At 12 months

- Support IT Operations Engineers in creating documentation for established processes and **become fluent in using our issue tracking system** and other internal SaaS applications; there is opportunity to take on administrative roles in the issue tracking/wiki software as well
- **Become an expert** in our internal user onboarding process, which includes provisioning user hardware, accounts, and access to various systems

» **Prioritize and schedule** helpdesk issues, provide post resolution follow up, and identify issue trends while communicating appropriate information to the team
» Collaborate and contribute to larger projects with our amazing team of Regional Leads and the IT Operations Engineers

What You'll Need to Be Successful

» 2–4 years or more of customer service, software support, and/or technical support experience
» A **passion for technology** with **attention to detail** and a combined interest in big strategic goals with exceptional time management skills
» Excellent interpersonal, written communication skills with the enthusiasm to learn and assist the entire team
» A **do-whatever-it-takes attitude**
» An affinity for a strong company culture with high-performing teammates who care about making a huge impact while maintaining the global user experience at MuleSoft

From Studio Artist to IT Wizard: Wayne's Take on IT

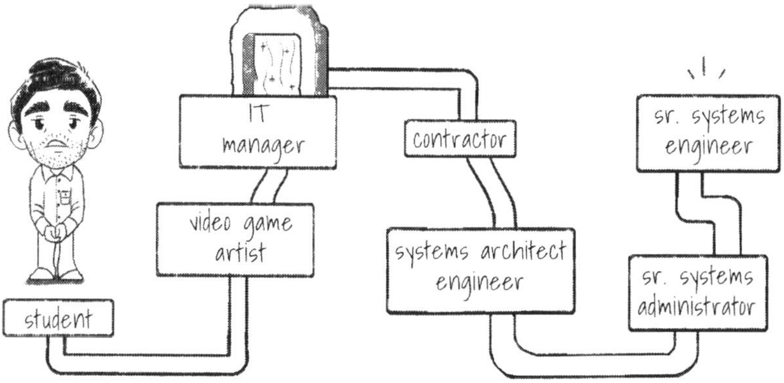

THE NUMBER ONE BENEFIT OF INFORMATION TECHNOLOGY (IT) IS THAT IT EMPOWERS PEOPLE TO DO WHAT THEY WANT TO DO. IT LETS PEOPLE BE CREATIVE. IT LETS PEOPLE BE PRODUCTIVE. IT LETS PE-OPLE LEARN THINGS THEY DIDN'T' THINK THEY COULD LEARN BEFORE, AND SO IN A SENSE IT IS ALL ABOUT POTENTIAL.

— STEVE BALMER, OWNER OF LA CLIPPERS, PREVIOUS MICROSOFT CEO

Most IT people break in by doing the helpdesk support or facilities type work that involve day-to-day tasks like setting up phones and computers. That's how you get the foundation of IT 101 and learn how to solve problems. As you grow, you tend to learn more specialized skills that could involve servers, networking, and switching, security, applications, etc.

The more specialized, the more valuable you become. However, don't take my word for it. Here's an interview with Wayne who started as an artist and moved into the gaming industry before going into the IT side at Salesforce.

Irwin: Tell me a little about yourself.

Wayne: I have a bachelor of arts degree in studio art, which is basically taking photography classes, learning some art history, and doing some drawing and painting. When I graduated, I found there was really not much of a career path unless I wanted to draw or paint. The university I went to, William Paterson University in New Jersey, had a program called a Master's Degree in Computer Art and Animation. From there I learned how to do art on the computer. We had these systems called Silicon Graphics systems, which back then were the only systems that were used by Industrial Light and Magic and Pixar. They were the big video rendering systems for doing the graphics for all the big films like *Terminator 2* and *Star Wars*.

We were all just artists, and we didn't happen to have anybody that knew how to manage the systems because the systems ran on Linux, so it was a Unix-based operating system. The OS was actually called Irix, so somebody had to figure out how to set things up. How do we manage the operating system? How do we manage the network? I learned how to manage them in college.

Irwin: Where did you progress for there?

Wayne: I got my first job in Manhattan. I worked for architects doing architectural renderings and architectural walkthroughs. From there I was recruited to become a video game artist, and then I started working for 2K Interactive, 2K Sports, so I started working on the first 2K games.

We had a whole bunch of Silicon Graphics machines, but no one was really managing it, and they were going to hire somebody to manage it. I was not getting along too well with the art director at the time. The guy that was running the IT, it was a 200-person company, needed help, so he told me, "Help me for a few months. If you don't like it, you can always go back to doing art." I was like, "Sure. I'll try it out."

It turns out my boss quit at the same time our company, Visual Concepts, was bought by Sega. This was when the first Dreamcast was launched, so at that time, when the first Dreamcast was launched, my boss quit after E3, which was the first time 2K games were actually presented to the world. My boss quit, and the president of my company at the time became VP of development of Sega.

Since I was the only one working in the IT department, he told me, "We need to grow because we're going to become way bigger, so you're now managing IT," so that's how I got into IT. I started as an artist in '95, switched over in '98 into IT, and have been in IT ever since. I've worked for Restoration Hardware.

Then, I saw something very similar between art and IT ...

If you're a really talented artist, somebody that can draw really well, has a good skill set, and can develop a portfolio just by drawing, then you can get an art job. The same thing in IT. If you're a self-starter, learn how to program, and learn Python, then you don't have to go to college. If you just learn that engineering skill set, and I've seen that happen quite a bit actually where people just have an engineering knowledge, they can get in the door and become a programmer without having a college degree. In fact, when I was in the video game industry, a lot of the engineers did not have college degrees. They were just a bunch of young teenagers plugging away, trying to create games, and became engineers.

Irwin: You mentioned there's two different things. The language learned to become a programmer, and then there's the IT side where you're helping run the backend of IT.

What kind of skill set do you need? What do you need for both?

Wayne: For the backend: I was more of a Unix guy when I started into IT because I had that skill set. When I got into IT, I learned that. This was in the late '90s, so I had to learn how to build PCs from the ground up and motherboards, and it was really easy. At least I thought so. Let's say you work for Best Buy, you get to work as a Geek Squad person. I think you can get some of that skill set. It's not too difficult to self-start on that. You don't need a lot of money if you want to build a PC from scratch.

I don't know if today you need that skill so much. In the late '90s, it was really a big thing to build your own systems, make them a little cheaper, but now it's so easy to get commodity hardware that is already pre-built. That savings isn't there anymore of actually building your own.

Irwin: Tell me about how you got into Salesforce.

Wayne: I actually worked at Autodesk. It's kind of interesting how I got into Salesforce. I came in as a contractor on the operations side, and you just have to show initiative at a place. If you show initiative, it gets you really far.

I came in. I was working in operations, and I was just busting out tickets. Somebody told me, "Hey, Salesforce is great. Just get your foot in the door," and that's what I did. I got my foot in the door as a contractor. I was way senior than operations, and I had a much better skill set than an operations skills set when I did go to Salesforce.

I just came in, just busted my ass. I got tickets, IT helpdesk tickets, anything from creating user accounts to checking backups, making sure something is backed up, adding a file share, making sure the equipment is actually running. Next, I was on the engineering. Within three months of that, I was hired full-time, and then they moved me into the engineering team. What we do is almost on an architectural level, where we're designing systems.

I do the data protection for all of IT. We're looking at AWS, Iron Mountain, Google, Cloud, looking at a bunch of deduplication appliances to see how, instead of backing up locally, we can back up to the Cloud and move a lot of our workloads into the Cloud.

*

Wayne started as an artist. The opportunity came out when he had to solve a pain point, which was IT. From there he grew his career. He was willing to take a contractor role, which is key. A lot of people break into tech through the contractor opportunity. From there, it's a matter of showing initiative, adding value, and then joining or jumping to a company of your choice.

From Helpdesk to Cybersecurity Specialist: Josh's Take on IT and Cybersecurity

This second interview for the chapter is another two-birds-one-stone example. Josh is a cybersecurity researcher in the IT department, which is a HUGE trend in the WORLD and will be for the coming years. Cybersecurity has become a top priority item for businesses and governments, and the cybersecurity industry has negative unemployment, meaning more openings than people that can fill them.

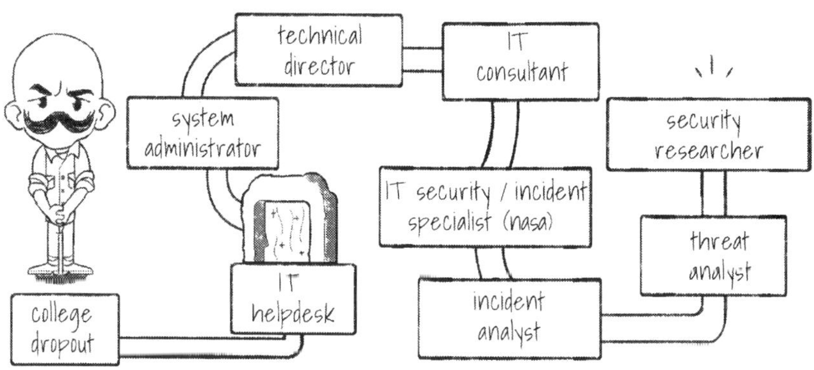

Josh is another person who doesn't have a college degree. I get excited about the second half of the interview because Josh talks about the many openings and high-paying roles in the security side of the business. He explains how he broke in and why cyber security is so important. Remember everything hinges on a company's ability to secure their customers' and their own data.

Josh has an incredible journey. He started playing with computers and got into tech working at an IT helpdesk. Next, he moved into doing some sales engineering where he supported account managers, empowering them to ultimately let them sell. From there, he worked at NASA where he met Natasha who told him about all the great things that we're doing here at OpenDNS (now Cisco Umbrella) and—boom! He becomes a security research person here, looking at threats (like WannaCry), helping create the algorithms used to block malicious newly created IP addresses, domains, ASNs, and more. For incredible insight into the IT security gateway, let's refer to Josh and his perspective.

Irwin: What advice would you give someone breaking into IT?

Josh: First of all, you have to be passionate about it. You're going to have a lot of things to learn. I read a lot of books, and I was building networks, so I could play video games with my roommates who were located down a very long hallway. I was using phone cables and things like that. But it was the practical, hands-on stuff that helped me because I didn't actually finish college. I didn't get a degree and then a job. I just started working, started basic menial jobs and then worked into tech like that. I had a passion for it. I kept following it and following it, and anything that I found interesting I learned about. I also had a loose vision of where I wanted to go. I studied towards that vision, and I continue to do that today.

Anyone who wants to break into tech, if you're coming from nowhere or even coming from a college degree background, you want to jump into it with both feet. If you've done nothing, start with tech support, do helpdesk, or help your parents and grandparents with their printers and computers. And then read up on why something works, why something doesn't work, about different operating systems, just learn how things connect together, why the internet works, how it works, and what the best practices are. That'll lead you into different paths of creativity and things to study that will lead you toward security.

While you're doing all that stuff, while you're fixing that printer, which doesn't seem like a security issue, just think about everything you're doing as leading towards that big vision. The more you know about all the base-level stuff, the better, faster, and more effective you can be when you're dealing with some kind of ransomware variant on the network or dealing with a hacker breaking in and stealing data. You'll be able to understand where they're stealing it from and how they do it.

Irwin: It's sounds to me you have to know the foundations of how IT works in order to defend against it.

Josh: Yes. So, I didn't finish college, but I do have a long history of going to community college. Work or life would get in the way quite often, and I actually got on academic probation once. City College of San Francisco kicked me out, saying, "You can come back in a year and see if you're better." I wasn't better after that, but I was really getting better at work, which was great. But I was terrible at school. I would take some courses, like Cisco courses at the city college. I took some courses at some other colleges. I didn't quite finish the Cisco course, but once I got to the point where I was like, "I'm tired of spending my entire Saturday here"—it was an eight-hour thing every Saturday—I just decided to start doing it on my own and skip going to the class. I didn't really care about the grade, which most people did care about. I cared about the experience. Also, I took some ethical hacking courses.

GOLDEN NUGGET

Here is another example of taking specific courses like Cisco & Ethical Hacking courses to build specialized knowledge to help him get started despite not finishing college & advancing his career!

Irwin: Ethical hacking courses? What's that about?

Josh: If you want to be able to catch hackers and bad things, you need to know how they do it, so you learn how to do it ethically. That means you don't go and hack some company; instead, you build a lab environment and you build it to be like what you'd like to hack and then you hack that. And you learn how to exploit vulnerabilities in software and achieve an objective. There's a whole industry of ethical hacking called penetration testing.

Irwin: I think you hit on a very good point: even though you're doing this manual stuff, you want to have a vision of where you want to go so that you're giving yourself purpose on the day-to-day mundane job of fixing printers or deploying laptops. What do you see as a trend in the industry?

Josh: There's a lot of trends, and there's a lot of marketing trends too. There's a lot you have to wade through. One of the reasons I go to a lot of security conferences—and I'm not talking about RSA-like big corporate ones, which I did go to as well—I go to DEF CON, which is a big one now. You can see people talking who are actual hackers and you can see where people's interests are heading. Usually, it's heading toward where the industry is really going. So, you can look at the marketing hype, everything, every article that you see or a billboard or anything you see in advertising tends to be machine learning, big data, things like that.

You've got to be able to adapt. That's actually the biggest lesson you learn in security is you have to adapt because things change. I mean if you're still trying to secure Windows NT or Windows 2000, you're not valuable.

Irwin: If I had to summarize our conversation today is, one, you want to have passion with computers. If you want to break into the IT side, which will lead you to the security side, passion is a big key, but the gateway to get in is through helpdesk, support, facilities operations, basically anything that requires getting the computer together. From there, if you want to create a future for yourself—and in your case, you were interested in security—the more you want to go there, the more you learn. You did your own self-development, your own growth by reading different articles.

Josh: Yes.

Irwin: You have to meet different people. You surround yourself and you learn about it, you get exposed to it. And you kept growing, right?

Josh: You've gotta keep growing, so I speak at conferences all the time. I come up with topics, research them, build a talk, and submit it to the conference. I get accepted fairly often, and this is what a lot of people do. You get accepted and then you can see all the other stuff and you can be a part of the community and you can give back.

Like you were saying in your summary, you have to be passionate about computers, and you also have to passionate about people because you're not just securing computers, you're actually securing the data and the lives of all the people that are managed by the computer. You want to meet people and learn how people work, so you can help them because you're securing the people.

*

As Josh suggests, there is so much opportunity in the security space, and it requires understanding how IT all works together. The huge importance of IT security is only increasing. Think about all the waves of businesses getting hacked and scared of the consequences if word gets out that they will be (or already have been) hacked—even if they didn't know it! If you like computers and may like to tinker on equipment versus write code, this is the gateway for you.

If a company does get hacked, it's important to have a legal structure to ensure that your interests are covered. When buying software or creating partnerships, there is always time spent drafting legal documents to ensure that both parties' risks are accounted for. The next chapter covers the legal gateway into tech, and you'll find there's even a place for people who don't have a law degree to enter as well.

CHAPTER 12

THE LEGAL GATEWAY

Are you interested in a career regarding law? Or did you pursue a law degree only to later realize that finishing law school and becoming an attorney isn't the right path for you? In a recent report less than half of recent law grads had 'good jobs' waiting after graduation. It's an extremely competitive field if you make a mistake during your career or you are graduating in a non-top 10 law school it may be tough landing on your feet. I was very curious to see if there is a way to break into the legal department without a law degree, and to my surprise, there is!

That's right! Today is your lucky day. If you have always wanted to work in the legal space and law school isn't an option or if you are tired of traditional law, it is possible for you to break into tech through the legal gateway.

Career Paths:

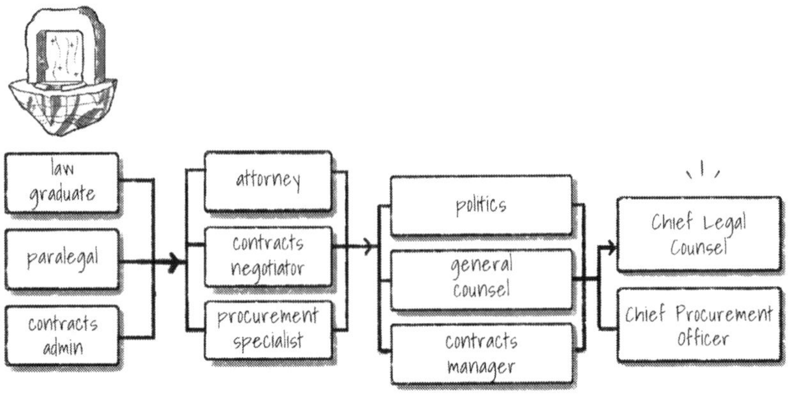

Gateways

» Contracts administrator/specialist ($55k–$105k)

» Contracts negotiator ($45-88k)

» Contracts manager ($72-133k)

» Paralegal ($38k–$71k)

» Procurement Specialist ($44k-$82k)

Skills

» Law degree is very helpful.

» Negotiation experience or training is helpful.

» Contract clauses and contract law (can be learned in pre-law online and onsite classes)

» Paralegal certificate or working as a paralegal is helpful.

Sample Contracts Specialist Description

One one the unknown gateways into legal is by being part of the contracts team! Here is a sample description for a Contracts Specialist!

Responsibilities:

- Entry and maintenance of contractual information in the Legal & Business Affairs department systems. Summarize and track SIEA agreements while maintaining consistent, accurate and relevant data.
- Outline contractual obligations and pertinent contractual terms; distribute fully signed agreements to various internal contacts.
- Fulfill internal contract approval and status inquiries.
- **Ensure** SIEA contract **compliance:** facilitate payment; request and provide certificates of insurance; complete product fulfillment obligations; notify internal contacts of time sensitive contractual obligations.
- **Test functionality** on new or existing legal databases prior to implementation.
- Aid in **developing training materials and user manuals**. Train non-technical internal business users on system functionality and associated legal processes and procedures.
- **Provide overflow support** to the Records Specialist in daily work during times of high volume.
- **Complete various ad hoc contract management requests and projects.**
- Provide verification of a company's legal name and business status using various government websites and online research. **Review of information and documentation** (e.g. company organizational documents, certificates of good standing, etc.) to **support verification. Work with in-house counsel** to resolve verification issues when they arise.

Requirements:

- Bachelor's degree or equivalent work experience.
- 2+ yrs. experience in the contracts field, contracts administration, or legal operations,

- **Must be able to understand complex agreements and legal terms.**
- Strong computer skills (must be familiar with Microsoft Word, Excel, PowerPoint, and database applications).
- Excellent verbal and written communication skills. Must be able to effectively articulate difficult ideas/concepts.
- Strong organizational, analytical and multi-tasking skills.
- Must be detail oriented and possess strong problem-solving skills.
- Must be **customer service oriented, self-starting,** able to **adhere to deadlines,** and work well with others.
- Interest in the gaming industry is a plus.

Before the Cisco acquisition of OpenDNS, I worked closely with Fallon, the lawyer featured in the upcoming interview, to ensure both my company and my customers' risks were considered, and concerns were addressed in the license agreement before the customer and our company solidified a partnership. It was incredibly impressive to me how our counsel team could decipher very complex language and address and add changes into a contract to ensure everyone's needs were met.

I had no idea that Fallon's experience as a contracts manager and counsel would lead to her being intimately involved in the legal process through the acquisition, especially considering she only had two years of experience in the area. This interview gives an inside look at the legal side of a startup. I'm super excited as this gateway has been a huge resource for me and my career, and I hope to see more people with law interests or backgrounds consider coming into tech. Plus, throughout the interview, you'll find golden advice from Fallon on how not to be intimidated by job titles and how quickly you can gain experience and move up in your career in fast-paced startups.

Entry into Tech via the Legal Gateway: Interview with Fallon

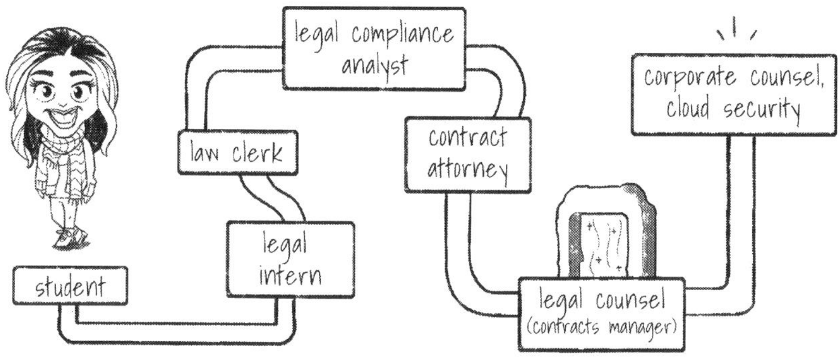

To help paint the picture of the day-to-day life of a contracts manager, here is my interview with Fallon. While going to law school and completing internships, she realized she did not want to take the traditional law path of working at a law firm. Check out our conversation to learn how this lawyer got into the tech world via the legal gateway.

Fallon: My story goes back to getting my undergrad degree. I knew I wanted to go into law. At that point, I had no idea what I wanted to do in the legal field. So, off to law school I went here in San Francisco at the University of San Francisco. I still didn't know what I wanted to do during my first year, so I looked at what my law school offered for certification. One was an internet and technology law certification. I took a few classes, which covered things like intellectual property, patents, trademarks, and copyrights.

In law school they teach a lot of these philosophical ideas and how to analyze things, but they don't necessarily teach the practical skills that you'll end up needing on the job.

I BUSTED A MIRROR
AND GOT SEVEN YEARS OF
BAD LUCK, BUT MY LAWYER
THINKS HE CAN GET ME FIVE.

— STEVEN WRIGHT

I quickly realized I didn't want to do the traditional law firm route when I interned at the district attorney's office. It was fun and exciting, but I really wanted to get into more of the business world. I didn't want to be a litigator; it required things that didn't things fit with my lifestyle, my personality. I wanted to get a better sense of the business world.

I graduated from law school in 2012, which was a very difficult job market. At that point, more than half of my graduating class ended up going into non-legal jobs because it was just so competitive. I ended up having to broaden my job search. I put all this work and money into law school, studying for the bar, passing the bar, but what was more important to me was finding a company that I liked, a business that I liked, and a role that I would enjoy. So, I found a contracts manager position at OpenDNS.

I went and interviewed. I was one of a few applicants that had a legal degree. It wasn't a requirement for this job, so the pay was not on par with what you would think an attorney would be making right out of law school, but I had the equity! So, I took the job, and I negotiated the title to be Contracts Counsel, instead of Contracts Manager, so I still got my attorney title. Part of the reason OpenDNS chose me is because I was able to take on even more than they were initially looking for sinceI was a licensed attorney. So that's how I got into OpenDNS.

Two years later we were acquired by Cisco. I was part of a handful of employees and executives that knew about the acquisition. I dove right in after 2.5 years on the job, essentially being told, "Hey, we're doing an acquisition. It's top secret. We have to continue to do our regular day-to-day jobs, but we also have to complete the diligence for this deal."

I basically didn't sleep for a solid six months, but it was a phenomenal experience because I got more experience in that one six-month period at OpenDNS than a lot of junior attorneys get at big law firms in five or ten years.

Now I'm part of a huge legal team at Cisco, which is awesome because I have all of these colleagues whom I can learn more from and bounce ideas off of. I'm getting to dabble in other parts of the business, instead of just what I was doing at OpenDNS.

Irwin: What are the differences between working in legal in a startup versus in a larger corporation?

Fallon: There are a lot of differences. I think one of the biggest differences is that it's a little more scrappy in a startup environment. The sales team really drives a company. If we don't have the sales, we don't have the numbers, and the company doesn't exist. It's a hard-lined kind of balance between, "Okay, we're sales-driven because we need to be," but at the same time you need to protect the company legally and not take on too much risk and scare off our investors or potential acquirers in the future.

The major difference is, back when we were a little startup, we didn't have that much leverage in negotiating agreements with bigger-named customers. You'd have to really make sure the business was comfortable taking on bigger risks that a company like Cisco might not have to take on because they have a proven track record of business systems and policies that companies tend to be more comfortable with.

Also, when you're in a startup, you really have to be prepared to do a lot of your own research and to use outside resources. It's really important, and especially in my case, if you're a team of one, to have outside mentors that you can go to. I relied on outside counsel for overflow support and for a lot of our corporate matters.

It's really important in that smaller stage to make sure that you're going the extra mile to learn as much as you can, as quickly as you can, because there is no one that's going to just teach you. They're just going to say, "Figure this out," and you've got to figure it out. Whereas in a corporation, they've already done a lot of this, so you have these processes that are already in place that you can just rely on, like "Okay, here's the playbook for that. Here's the template for that. Someone's already done this. I'll just ask them."

There are pros and cons. You can get a little more creative at a startup than you can at a public company.

Irwin: For those who may be going through law school right now or are looking to go into it, what was it specifically that you liked most about coming to the business side in the technology field?

Fallon: I love my legal colleagues at Cisco, but I also really enjoy working with non-legal individuals. If you're interested at all in entering the startup world, like technology, it's important to get a feel for other industries, other departments, other roles that you're simply not going to get at a law firm, even if you're at a law firm that is representing technology companies.

Irwin: What's fun about working at a non-legal company?

Fallon: The products. It's the founder's product and baby, but it also becomes everybody's baby. Whereas if you're going a traditional law firm route, you have a lot of different clients. When you're going into technology, as a startup, that's your client. It is that company, and it is that product, and you're doing everything you can to protect that company and that product, and make them successful.

Being a part of a startup, a lot of times you take on work that's not necessarily in your role because someone has to do it. That's really a big consideration when you're deciding which route to take when you're in law school.
In a startup, the company wants to get the sales deal done, and of course, both sides of a negotiation are lawyering up. As the lawyer, you're doing everything you can to help close the deal, but you're doing it in a business-friendly way as much as possible. In any contract negotiations or sensitive legal topics it can still, of course, get a bit heated, but you've got to have the personality that can do both and know when it's appropriate to do both too. I think it's fun navigating those waters.

Irwin: Is there anything else other than contract manager that you see that doesn't need a legal degree?

Fallon: You could always be a contract negotiator or be a contract manager, anything to that extent. What a contract negotiator does, is work with the business, figure out what risks the business is comfortable with, and then advise the business on what risks they think they should or should not be comfortable with. At the end of the day, it's a business decision though, and you just negotiate the purchase or vendor contracts. You don't need a legal degree to do that. You do need some sort of understanding of contract clauses and contract law, but you can do that in pre-law classes that you can in undergrad or even earn a paralegal certificate or start as a paralegal.

That is a much easier process than law school. There are also online courses. There are so many courses out there. It sounds so cheesy, but even in my early days, I just took as many classes as I possibly could to learn new ways of negotiating. New styles of negotiating. You know, compromises that you wouldn't otherwise think of, and that's something that anyone can do. You don't need a legal degree to take an online course like that or go to a conference.

Irwin: Any advice you would give to someone graduating from college and looking to break into tech?

Fallon: Don't be afraid of job titles that aren't exactly what you want. That's my biggest piece of advice. That's how I found my job, that's why I have my dream job today, and I have so much experience in such a short time, so don't be afraid of job titles. You can always negotiate.

Irwin: Are there any questions I should be asking that I haven't asked yet?

GOLDEN NUGGET

Fallon: If you're really just starting out of undergrad or even law school, if an industry is really important to you to be in or a type of job is really important to you to be in, don't be scared of initial salary because if you're a rock star and you prove yourself, in a startup world it can change that quickly and easily for you. And it will because you've become valuable to them, but they want you to prove yourself first. Choose what you want to do first. I did that and it paid off hugely for me and I'm loving what I'm doing. I'm so lucky to now be at a company like Cisco on their legal team.

*

This interview is golden! Fallon adds really good advice around starting out. **Don't be afraid of job titles that aren't exactly what you want.** When starting out, if the industry and type of career is important to you, don't be scared of the initial salary. If you are a rock star, your value will grow and great leaders will make it happen to ensure you are happy (right role, compensation, etc.) at the organization.

So, at this point, the conclusion of section 2, you've encountered all the gateways. You probably have one in mind as the one you'll pursue to break into tech. The next section, The How: Putting Words to Action, will lay out all the behind-the-scenes prep you'll be doing to land that position in your chosen gateway.

Now, time for section 3. Let's get moving!

GET YOUR FREE BONUS!

Thank you for reading & sharing this book!
To help with your **momentum!** Here are **4 FREE videos** I've included for you to bring the **book to LIFE!**

Go To: www.irwinki.com/p/FREETRAINING

If you want to see more videos!
Follow me: Search "Irwin Ki" on Youtube

Find me via @Irwin_Ki on Instagram

Get **connected with other readers** via FB group:
"Breaking Into Your Silicon Valley Community"

SECTION THREE

THE HOW

PUTTING WORDS TO ACTION

CHAPTER 13

FORGING YOUR MAP

In sections 1 and 2 you were gathering information and educating yourself on the often unknown entry points into tech careers: the gateways into tech that don't require you to have a CS or math background. You even learned that some of these gateways don't even require a college degree! Now it's time for acting, moving, doing, performing, executing. So, without further ado, welcome to section 3—The How: Putting Words to Action. In this section I've compiled stories, lessons, templates, and recommendations, so you can get to take action and actually advance into your tech career via your chosen gateway. Section 3 aims to help you shortcut your tech career search, so you can enjoy the best chances when approaching your chosen tech gateway.

This section delivers a multi-faceted strategy so that you can not only stride across your chosen gateway into tech, but also so you can land the most lucrative offer.

We'll cover:

- » Your MAP
- » Your resume
- » Your LinkedIn profile
- » Your target list
- » Your interview
- » Your first offer

Remember, I've done a lot of hiring in my time. That means I've dealt with lots of resumes, evaluated many LinkedIn profiles, done a ton of interviews, and negotiated many first offers. I know what makes an applicant shine and what doesn't work well. We'll cover it all in this section, so that you can hit the ground running in your sprint through your chosen gateway into tech.

Surprising Himself, Defeating the Fear

So, you're poised on the edge, just one leap to get yourself moving through your chosen gateway—I know you are eager. But let's take a moment for one more story. One more critical story. This story is going out to my readers who still have doubt, who still think, "Maybe other people can break into tech through one of the gateways, but not me because I'm just not lucky." This story is for all of us who allow that destructive voice of doubt to quash our efforts at pursuing our dreams.

Several years ago, a friend of mine asked me a favor: "Irwin, my brother Andrew is a good guy. He needs a place to stay for a few months, but because he had to serve a few months in jail for misdemeanor marijuana charges, he doesn't pass a background check. He can't lease a place. Could he sublet from you? Be your roommate for a little while? He's good for the money and easy to live with."

I told her it was fine.

When Andrew moved in, he was still smoking a lot of weed. As I got to know him better, he also didn't seem to have purpose or belief in himself. Also, he was selling marijuana on the side to fund his habit and to have a little spending money.

One time, I asked, "Andrew, I bet you'd enjoy having a regular work schedule. Why don't you look for a job?"

Agitated, Andrew responded, "No one is going to hire me with my background."

I was taken aback. It wasn't like he had a felony, just a pending misdemeanor. Either way, I didn't buy it, so I challenged him to go with me to the local mall, see who was hiring, and start filling out applications.

Once we arrived there, it was evident that Andrew was very uncomfortable. Andrew was so afraid that he couldn't bring himself to go up to cashiers and ask for applications. So I did it, starting with an athletic shoe store. Then more stores. About every third store was hiring, and I got applications. I passed each one over to Andrew, and he started filling out the forms. I must have visited about 12 different stores and learned only five were hiring. By the end of the day Andrew had filled out those five applications.

Within days, Andrew was interviewed and then hired at a restaurant in the mall. He started work by the end of the week. At this job he was making more money and making it more consistently than when he'd been selling marijuana. He even made a new friend who worked at the local jewelry shop.

Even more importantly: after he started working, Andrew showed more purpose and happiness. His self-confidence increased. He started believing in himself and his future.

Andrew and Us

Andrew's story is our story. It is the story of each of us. Andrew allowed a fear in his head to distort his notion of reality. His fear kept him boxed into unhappiness, doubt, and more fear. That can happen to any of us and probably has happened to most of us in some form at some point in time.

Andrew's way out was through taking massive action. Taking massive action was fundamental. By doing so, he broke through and broke free from the cage in his head that his fear was holding him in. By taking massive action, he started a new direction in his life, one where he was so much happier and more confident.

The takeaway—don't let your doubts, fears, or limiting beliefs wall you in from pursuing your chosen gateway into tech. Sure, you can feel fear, that's normal, but you have to take massive action anyway. By doing so, you'll eventually leave that fear behind and embark on a whole new and interesting journey in your life. Just like Andrew, taking massive action might, at first, put you out of your comfort zone, but it's worth it considering the resulting happiness, success, and confidence you'll get access to.

It's no surprise too that taking massive action isn't easy. It takes time, effort, and energy. Know that and expect it. However, unlike Andrew who needed to be seriously pulled along (by me!) and then who employed a random approach, it's up to you to mobilize your inner self-starter and act on the recommendations I'm providing you with. I'm guiding you, but I'm not available (metaphorically) to go up to the cashiers for you, get the applications for you, and make you fill them out. You are going to have to call on your inner self-starter to get yourself acting on the guidance I provide. What's great, though, is that my guidance is very specific; there's no arbitrariness in this approach. Your approach to breaking into tech will be highly strategic, so the time, effort, and energy you expend will be done very efficiently. With that said, it's time for your first action.

The first massive action you are going to take is creating a map for yourself. Well, really it's a MAP, a "mutual action plan."

Your MAP

Before diving into your MAP, let's determine your timeframe. For those in my sector of tech, which is "enterprise software sales," we live quarter-to-quarter, so that's three months or 90 days at a time. When I start a new sales role, my first goal is to create my first win, no matter how small or large, within my first quarter (or 90 days). That's what I'm asking of you as well: that you aim to achieve your win in 90 days. What's your win? Is it landing a job in your chosen gateway into tech or is it getting your first onsite interview? (I challenge you to choose the former but create your WIN!)

As we learned from Andrew, it's all about taking massive action, even in the face of any fear and doubt. So, let's get you on the road to massive action.

To make this even more official. I recommend reading this contract, ripping it out of the book, sign, and post this somewhere you can see it every day.

THE PATH TO SUCCESS IS TO TAKE MASSIVE, DETERMINED ACTIONS.

— TONY ROBBINS

GOLDEN NUGGET

Those who write down their goals EVERY DAY are 42% more likely to achieve it."

Here's what I want you to commit to:

I, [first name], am committed to landing my first tech job in the [your chosen gateway into tech] gateway by [give date].

Signature: ....

Now, onto the rest of your MAP, starting with my first encounter with a mutual action plan (MAP). In technology sales, the product I sell affects multiple departments in the business I'm selling it to; thus the sale will involve multiple decision-makers in that business. I got a glimpse at a large number of decision-makers and the complexity behind their coming to a decision about buying a product from me when they included me in one of their group emails. The subject of the email read, "Go/NoGo MAP for ACME Corp.—pls review MAP." When I read this, I remember wondering, "What's a MAP?"

To find out, I opened the email and then took a look at the Excel document that was attached, which was the MAP. The Excel document gave a table that laid out the actions, goals, completion dates, and responsibilities of the various decision-makers involved. It looked sort of like this:

Action	GOAL	Dates	Responsibility
Jack to confirm with Fred that Tim (economic buyer) is set for 1/27 meeting	Confirmed	1/25	Jack (Champion–ACME)
ACME+BMC presentation	Win over Tim (economic decision-maker/buyer)	1/27	Fred (AM–BMC), Jordan (SE–BMC), Adam (VP–BMC)
ACME's go/no go on solution	Understand procurement next steps and timeline if no/go understand objections.	1/28	Jack and Fred
Understand procurement process if it's a go	Understand unknowns (how long, redline process, who signs final doc) and plan for target close date	tbd	Fred (AM–BMC) and Adam (VP–BMC)
CFO or Todd (economic Buyer) (Need to know who that is to sign off)	Signature by 3/30	GOAL: 3/30 end of quarter	Final signer and Fred (AM–BMC)

Then my eyes opened up! This mutual action plan (MAP) was a fabulous tool because it clearly laid out each step, showing who was involved in a particular step, the targeted completion dates for the particular step, and how all of that worked with the bigger goal and final deadline for making the decision. If the account manager (AM) could drive these steps, there was a clear path towards closure at a specific deadline. Needless to say, I was in awe of how a MAP made this potentially complex process streamlined and simple.

I want you to have a similarly streamlined, elegant, and easy-to-use MAP to put to use in your pursuit of breaking into tech via your chosen gateway. I created a MAP that takes into account and neatly sequences the various steps you're going to be taking to reach your ultimate goal. This MAP will help you neatly organize the many actions, goals, completion dates, and responsible parties. Also, I've coordinated the chapter of the book associated with the given action and goal, so you can quickly refer to that chapter to refresh yourself on details. With all this mapped out in a single table, you are best equipped to meet your big goal by 90 days' time.

Here are the 10 steps needed to break into ANY gateway job. Here is your MAP.

ACTION	DO IT	GOAL
1. Choose your gateway and start thinking about WHY you want to break in.	Completion date? ________	Get clear on your WHY and WHERE to focus
2. REVAMP your Resume (see Irwin's resume toolkit) and LinkedIn.	Completion date? ________	Simplify your resume w/ metrics and create a LinkedIn account that is findable.
3. Determine where in the world you want to work and the size of the company you want to work at.	Completion date? ________	Understand where the opportunity lies and what kind of company fits your needs.
4. Build your target list with gateway roles, decision-makers, and referral peers.	Completion date? ________	Create an attack plan and understand the people who can help you out.
5. Research and get in contact with peers for coffee, quick call/meeting, etc.	Completion date? ________	Find different ways to get information by read study, get jobs to become that person or get experience to help get into the gateway job.
6. Take massive action applying online	Completion date? ________	Play the numbers game and get in front of the decision-makers.
7. Reach out to decision-makers.	Completion date? ________	Create other ways to break in
8. Reach out to external recruiters.	Completion date? ________	Add additional pipeline of interviews.
9. Crush interviews.	Completion date? ________	Practice with a friend or family member so your 1st interview is not your 1st. Be their ideal candidate.
10. Negotiate multiple offers.	Completion date? ________	Have choices to make the best decision.

I urge you not to simply fill in the MAP but actually use it. Refer to it regularly. Use it as a resource. Don't glance at it and then neglect to view it again. Print out copies of it and post them, so you can see them regularly. If you put this MAP to use, it will organize, motivate, and push you to take that massive action, which plays out over many, many steps.

So, like Andrew, you are going to take massive action. That's what section 3 is all about. By committing to achieving your goal in 90 days time, you've empowered yourself by having a distinct time frame during which you'll be acting. With the MAP I designed for you, your approach will be strategic and efficient (i.e., nothing arbitrary here!) to ensure you break into tech via your chosen gateway by that 90-day deadline.

Now that the clock is running, let's get moving. In the next chapter, you'll be taking massive action in terms of strategically crafting and adjusting your resume and your LinkedIn profile, so they will make you shine for your chosen gateway into tech.

CHAPTER 14

CRAFTING YOUR RESUME

AND APPLYING IT TO YOUR LINKEDIN PROFILE

Before a hiring manager or recruiter even sends you a reply email, lifts the phone to call you, or is willing to let you in the door for a preliminary interview, there are things they look at about you that you MUST have in tip-top shape. Getting these right is super important to get your foot in the door. Here's what they look at in the order they look at it:

1. Your resume (CV)
2. LinkedIn profile (sometimes this is the first thing!)
3. Social media profiles: Facebook, Twitter, and Instagram (clean this up!)
4. What shows up when they google your name.
5. Any back-door reference (people they know that is connected to you)

In this chapter, we're going to address that first step: your resume and LinkedIn profile. It's on you to get numbers 3 and 4 spot-on. And number 5, we'll address that in chapter 17. Let's jump into your resume.

The Sterling Resume

As a hiring manager, I've evaluated many resumes. Below you'll find what a common zero-tech-experience resume looks like. Notice, I said "common," which doesn't mean it's awful, but it also isn't superb (which is what yours is going to be). To make yours superb, you must first avoid the common mistakes. When examining this "common" resume, these are the trouble-spots that I want you to notice:

- » Mistake #1: Poor formatting
- » Mistake #2: Typos and grammar mistakes!
- » Mistake #3: Fluff and generalizing (few specifics)
- » Mistake #4: Lack of growth and continued education

Mistake #1: Poor Formatting

I'm a big believer that 80% of resume issues can be solved by mirroring a successful resume template. Most resumes look like dookie as they are usually pulled from the first template found in a Google search. Here's the shortcut ...

> ### Solution #1: Time Tested Template
> I can't recommend it enough that you transfer your resume to a tried-and-tested template . Go to **irwinki.com** to find the template that my friend used when applying to and landing a job at Google. Also, it has helped me get into BMC, OKTA, and OpenDNS (now part of Cisco).

I remember my first time transferring to this template. I was shocked how professional I felt with the proper formatting. Looks matter. This template will prevent common mistakes because it will:

» Make your resume ATS (applicant tracking system) friendly
» Keep your resume concise and to the point
» Force your resume to be only one page long (you may need to consolidate similar experiences)
» Require you to lead the resume with your accomplishments

Mistake #2: Typos and Grammar Mistakes

More than half submitted resumes, 58% to be precise, have typos!

Solution #2: Grammarly and friends

Run the resume through Grammarly. Grammarly is an enhanced spell-check software. Use it on your resume. Also, proofread it several times yourself and ask others to look it over. The more sets of eyes, the better.

Mistake #3: Fluff and Generalizing (Few Specifics)

Most resumes describe what the person did for each job listed as **opposed** to highlighting particular skills gained and accomplishments achieved for each job that are relevant to the job the person is applying. Resumes like this end up getting glossed over. Here's a brief example of a generic, fluffy resume:

Summary

A hard-working and enthusiastic employee with exceptional customer service and interpersonal skills. Experienced in office, retail, and customer service settings. Able to learn quickly and manage simultaneous assignments and responsibilities, as demonstrated by successfully pursuing higher education and work.

Fine Dining Server/Hospitality, 2016-Present
Palomino, San Francisco, CA
Greet customers, serve cocktails and food. Assist with food running and bussing tables. Assist guests with all of their needs and provide excellent customer service. Extensive wine and alcohol knowledge acquired throughout duration of employment.

Solution #3: Metrics and Case Examples

Give metrics and use numbers to cut through the fluff and get specific. Metrics are the numbers behind your work.

Here's that "Fine Dining Server/Hospitality, 2016–Present" example from above after metrics have been applied to it:

> Server, Palomino San Francisco, CA   2017–Present
> Assist guests in creating an optimal fine dining experience
> Average $3,000 personal net sales a night
> 1 out of every 4 tables engaged in upsells that increased their check average by 25%

BONUS TIPS!

Use these impressive buzz words: *achieved, managed, resolved, launched, increased, engaged*

Mistake #4: Lack of Growth and Continued Education

I love this quote from Naval Ravikant, the CEO of AngelList: "Free education is abundant, all over the internet. It's the desire to learn that's scarce." Continual growth is important if you want to be competitive and relevant in an industry that is constantly changing. There are two types of growth. The first is growth based on interests, like picking up a new hobby, like krav maga, painting, or ballroom dancing. The second is growth tied to your professional life. This could include online certifications, Toastmasters, business courses, and books tied to your industry... It's important to have both, but if you want to get a leg up on the competition, focus on professional growth.

Solution #4: Find Your Gaps and Learn!

Each gateway offers clues on the skills, certifications, or strengths required. Make an effort to better yourself by filling in those gaps. By doing so, you will significantly differentiate yourself from other candidates. There is a lot of online content and courses today that are available for streaming/downloading at very little cost. As a matter of fact, we've discussed some of these at several places over the course of this book! It's a lot like Neo in The Matrix: you just need to download that data into your brain and you will be stronger, smarter, and faster. There are no shortcuts ... but if you are willing to learn, your options will increase many times over.

DON'T WAIT FOR OPPORTUNITY,
CREATE IT.

— ALLCUPATION.COM

The Best Zero-Tech Experience Resumes for Breaking into Tech

The best way to learn from the above "common" resume is to revisit it after it's been revamped. We're going to revamp it following these seven steps:

Step 1: Transfer your resume to a tried-and-tested template. Get yours at **www.irwinki.com.**

Step 2: Reflect back. Instead of describing what you did, get creative and change it into metrics of what you accomplished. Lead with any honors, awards, and accomplishments you are proud of.

Using metrics allows you to bypass the most common mistakes, which are the inclusion of objective statements, stories, redundancies, and unnecessary words.

Step 3: If your resume suggests you've been job hopping (2–8 month stints at jobs), then change it to years.

Step 4: Feature yourself in your ideal future role, for example, "Aspiring Customer Support Specialist" and mention the qualities that you WILL bring to the table that are most relevant to that role. You need to start visualizing and transforming yourself into that future professional you seek to become.

Step 5: In the education section, do not put your graduation year. This helps you avoid the possibility of age bias. Removing your address information as well is a good idea to remove location bias as well.

Step 6: Start reading books, taking online courses, and getting certifications to add differentiating skills to your resume. Check out Udacity.com or any resources mentioned in the gateway chapters, and I'll post recommended resources on www.irwinki.com.

Step 7: Do a spell and grammar check via Grammarly, but also human eyes and brains are needed! Get proofreading assistance from friends, family, or a professional.

GOAL: Keep it simple and wow the hiring manager with your metrics in your listed non-tech jobs to show all that you accomplished (which is different than telling what you did). You need to properly flex who you are today and who you are becoming.

When this is done, review it with a mentor and be proud. Congrats! You've taken steps forward in your career.

Get on LinkedIn!

Linkedin is the leading Facebook for professionals. It is a MUST for technology careers. Recruiters source and qualify based on your profiles and the words that show up on your profile. The challenge for newbies is most searches are based on brand-name companies, so it doesn't work so well if you are new or in a startup that no one has heard of. Don't let that discourage you. The goal of crafting an outstanding LinkedIn profile is, again, to reflect your metrics (what you've accomplished, not what you've done on a job) and where you are heading.

Steps for an Effective LinkedIn Profile

- **Step 1:** Change your Linkedin headline and most recent title to "Aspiring ____ [gateway title]." For example, "Aspiring Sales Development Representative (SDR)"
- **Step 2:** Transform your job experience on LinkedIn by including those metrics and accomplishments you highlighted in your revamped resume.
- **Step 3:** Take a quality photo. Get dressed up and put your game face on. The new iPhone and Samsung phones have great cameras. Also, Sephora will do your makeup for you for free.
- **Step 4:** Connect your email and LinkedIn mobile on your phone and add your contacts. This allows you to see expand your 1st, 2nd, and 3rd-degree connections. Think mutual friends on Facebook. If you have a friend that works at the company you are applying for congrats that's a 1st-degree connection. Hopefully, he or she can help refer you in. Chances are you may have more 2nd and 3rd-degree connections. These are your connections who have friends or connections that are at the company. By being resourceful you can reach out to your connect who to help you connect you to their friend in hopes for referral, introduction, or a just a friendly nudge that you are interested in working at that company.
- **Step 5:** Leverage your alumni network. There are alumni groups on Facebook and LinkedIn and you can get connected with the community that you already have!
- **Step 6:** Leverage the power of social media. By posting "Hi Facebook Friends! Do you know anyone working in the technology space? I am determined to work in that growing industry. If you don't mind tagging someone you think who can help it'll be greatly appreciated." All you need is that one connection that can change

the future of your career direction and having an insider helps! There's a chance that you already have that in your network. You just have to ASK.

BONUS TIPS!

A. Check out facebook group: "Breaking Into YOUR Silicon Valley Community" to connect with other gateway seekers to find an accountability buddy to share resumes and practice interviewing. It's great way to ask me questions and connect with a peer at a similar stage in life and help each other during the MAP process!!!
B. Albert's list is another great facebook group as well.
C. Under the summary section you can share your story and even create a Youtube and LinkedIn video of yourself selling yourself as the next A player. It can help!

Proof in the Pudding

Leah is a server at a local neighborhood restaurant. I've been helping her do these exact same steps. Once she changed her LinkedIn headline to "Aspiring SDR," recruiters have been reaching out to her nonstop! This one change led to her receiving messages from more than +20 interested recruiters. Why? She's made her intention known and easily searchable.

Here's an example of this simple change:

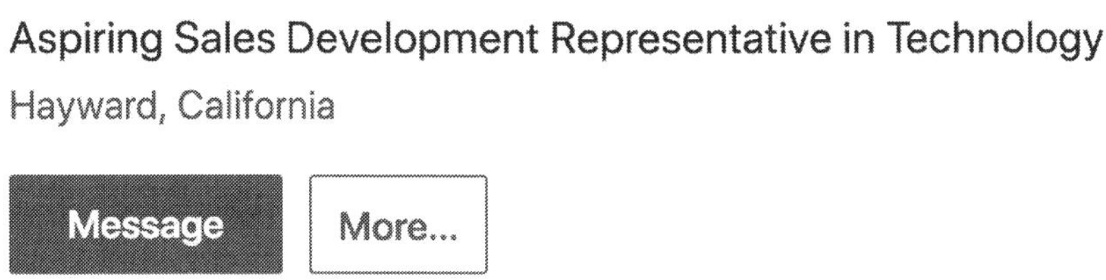

This small change, "Aspiring [gateway title]," is one of the fastest ways to get some traction! Remember, these gateway roles are always going to be in-demand. The easier you can make yourself searchable the better!

Now you have it, all the steps for making your LinkedIn profile stellar and effective. First revamp your resume, following those given steps and recs and then work on your LinkedIn. Make sure you allot in your MAP the appropriate number of days required to transform your resume from "common" to "amazing." As I told you at the beginning of section 3, it's all about taking massive action, even in the fact of any niggling doubts and fears. That's how you'll break into your chosen tech gateway and do so in our 90-day timeframe.

And your massive action is only just beginning. Our next chapter addresses your next action task: pinpointing your where. As you'll soon learn, not all locations are equal when pursuing your tech career. We'll explore the location issue in depth so that you have plenty to go on when you determine your where.

CHAPTER 15

PINPOINTING YOUR WHERE

TECHNOLOGY, TIMING, TALENT AND TERRITORY

In my business (tech sales), success can be summarized in the these 4 Ts—technology, timing, talent, and territory. In section 1, we explored the benefits and intricacies of the many gateways in the world of **TECHNOLOGY.** In the previous chapter we mapped out the end result & desired timing by breaking minor actions to **MAP** your overall goal. This helps create a helpful structure for your **TIMING.** In the previous chapter we covered how to build your resume and the importance of continued growth to help showcase and improve your **TALENT.** The next most important piece is your **TERRITORY.**

Territory-Related Tech Trends

I urge you to open up your mind to a tech reality: knowing *where* you want to work is as important as determining *what* companies you are interested in working for. Three-quarters of US venture funding went to three states California, New York, and Massachusetts. What this means is there are certain locations where you'll find way more tech companies. If you happen to live in a place where few tech companies are located, that's going to affect (and probably limit) your options for the tech companies you can realistically pursue working for.

The location factor is something tech reporters have seriously researched. In *The Four: The Hidden DNA of Amazon, Apple, Facebook, and Google* author Scott Galloway narrows the location factor far beyond the three states that received the most venture funding in 2016. In this excerpt notice Galloway's argument about location and tech:

> For years, we believed the digital age would enable us to "work anywhere"—a utopia of people living in quiet mountain cabins, tapping away at their laptops through the magic of the information superhighway. In fact, the opposite has happened. Wealth, information, power, and opportunities have concentrated, as innovation is a function of ideas having sex. Progress is typically in person. Also, we are hunter-gatherers and are happiest and most productive when in the company of others and in motion.
>
> More than 80 percent of the world's GDP is generated in cities, and 72 percent of cities outperform their own countries in growth. Every year, a greater percentage of GDP moves to cities, and it will continue to do so. Thirty-six of the hundred largest economies in the world are US metropolitan areas, and in 2012, 92 percent of jobs created and 89 percent of GDP growth came from those same cities. And

> not all cities are equal—the global economic capitals are becoming supercities. New York and London consistently rank as some of, if not the, most powerful cities in the world. Developers are also keen to invest in wealthier cities, where they can expand accordingly (think of Manhattan businesses that are expanding to Brooklyn locations). It appears that the lottery economy applies to real estate, too.

Based on Scott Galloway's findings, it is specific cities and "supercities" that host the most career opportunities. In *Angel: How to Invest in Technology Startups* author Jason Calacanis narrows this location argument even more. In this excerpt you'll see that Calacanis provides ample evidence to show that, as given in the title of the chapter the excerpt comes from: "Silicon Valley Is the Center of the Universe":

> Over the past five decades, Silicon Valley has emerged as the driving force not only in technology but also in media, transportation, advertising, health, and lodging. No region in the world drives more societal change right now than the tiny peninsula in Northern California that is made up of cities like Mountain View (a town synonymous with Google and LinkedIn), Palo Alto (Facebook), San Jose (eBay, PayPal, Cisco), Cupertino (Apple), and San Francisco (Twitter, Uber, Airbnb).
>
> If you look at the top five companies in the world by market cap right now, they are all technology companies: Apple, Google, Microsoft, Facebook, and Amazon. Of those, three are located within fifteen miles of one another in Silicon Valley and the other two are in Seattle. Of all the venture capital money being invested in the United States, 30 percent is invested in the Bay Area. That's nearly double the money that's invested in Boston, New York, and Los Angeles—combined.

> The Bay Area is the center of the world—it's not even up for discussion. And I'm not just talking about the center of the startup world, I'm talking about the entire world. If you look at how transportation is changing, it's being completely reinvented by Tesla's electric vehicles, Uber's ride-sharing network, and Google's self-driving cars. If you look at lodging, the most disruptive force is Airbnb. If you look at politics, Twitter and Facebook are driving the elections—not DC or newspapers.
>
> If you look at advertising, Google and Facebook are making the lion's share of the most important advances with things like targeting and retargeting of consumers. That's a space where newspapers, which have lost 70 percent of their revenue since their peak in 2000, are no longer leading. In fact, more than 50 percent of mobile advertising money is spent with Google and Facebook—and mobile isn't the future anymore. It's everything.
>
> In entertainment, the most important company isn't in Los Angeles. It's Netflix, located in Los Gatos, a town of thirty thousand people formerly known for its orchards. Netflix now has more than eighty million subscribers and is spending more than $1 billion on original content yearly. In just three years, Netflix has caught up to HBO's level of quality with shows like House of Cards and Orange Is the New Black. That's even more amazing when you consider that HBO started its Emmy-dominating original programming efforts back in the '90s. All this from a company that specialized in mailing DVDs to your house in big red envelopes just a decade ago.

Based on Calacanis' experience, you are correct in thinking to move to the Silicon Valley area to enjoy the most options for starting your career in tech.

But, I also recognize that this isn't right or possible for all my readers, so I'll be offering you lots of advice and information to help you pinpoint the location that's best for you in your pursuit of your tech career. With that said, let's step away from Silicon Valley for a moment, and consider the whole world.

The good news is there are technology hubs popping up all over the world. From the chart below, featured in a 2018 Bloomberg article, notice the vast range of cities receiving venture funding. The first four are the expected ones, but beyond that, notice that the American South and Midwest also host tech cities.

Think international as well. The largest technology companies are coming out of China. Ever heard of Alibaba, Tencent, HUAWEI? They are HUGE! Sweden has been a hot spot for tech as well. They've produced Spotify, King (Candy Crush), Mojang, Minecraft, Skype, and SoundCloud!

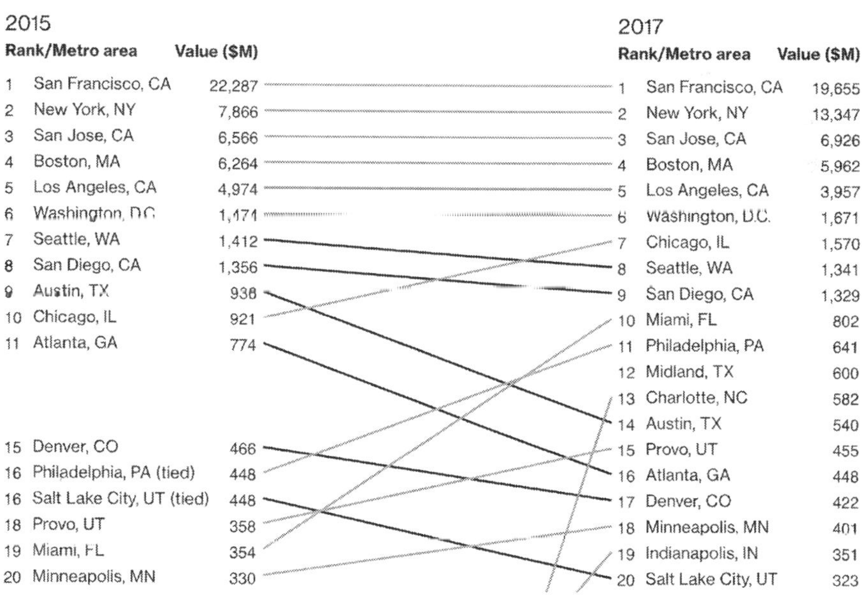

A DECENT PROXY FOR A
TWENTY-SOMETHING'S SUCCESS
WILL BE THEIR GEOGRAPHIC TRAJECTORY.
HOW LONG DID IT TAKE THEM TO GET TO
THE BIGGEST CITY IN THEIR COUNTRY, THEN
TO THE BIGGEST ON THE CONTINENT?
THE STRONGEST SIGNAL OF SUCCESS WILL
LIKELY BE THOSE WHO MOVED TO GLOBAL
ECONOMIC CAPITALS, THE SUPERCITIES,
VERSUS THOSE WHO STAYED IN THE
RELATIVE HINTERLANDS.

— SCOTT GALLOWAY IN THE FOUR

When US startups are looking to expand globally, the first office is usually London or Dublin for European coverage, followed by Sydney for Australia and Asia, then Dubai for the Middle East. The point is software really is eating the world.

With all this said, before you start building a list of tech companies that you want to apply to, you need to ask yourself, "Where in the world do I want to work?"

On the one hand, tech companies are popping in many locations all over the world; on the other hand, most startups that get funded and exited are located in Silicon Valley. If you want to put yourself closest to the action, where the opportunities ARE endless, then you should be willing to relocate to or work for a Silicon Valley tech company.

If you are in a rural part of the world, you must ask yourself whether you are willing to move for a better opportunity. This is super important and very American. The Pilgrims moved for their religious freedom, early Americans went West in search of gold, my parents moved to the United States from Hong Kong and Malaysia for better education and overall opportunities. In all these instances, it wasn't easy, comfortable, or seamless to relocate.

Taking major action despite your fears can lead you to finding greater happiness, fulfillment, purpose, and opportunities. What about you—are you willing to choose opportunity and growth over comfort and familiarity?

Are you willing to move to a city? If yes, then you will have more choices and opportunities. If not, the choices available in a non-urban area will be limited, and equity upside will be likely non-existent. Let's look more at tech careers in non-urban locations.

Options in Non-Urban Locations

If you want to work in technology but not move to an urban area, I recommend that you focus on the IT and developer gateways, or breaking in via a local technology partner or tech recruiting firm.

IT and Developer gateways—every company needs technology, and most corporations, non-profits, and educational institutions have IT teams. Some companies may have needs that will require development. A lot of time this development is outsourced, or the software is purchased from the manufacturer directly. Either way, IT and developer skills both will help if you want to parlay it into another technology company in the future.

A local tech partner—if you don't want to pursue those two gateways, then start off at a local technology partner. Value-added resellers (VARS) and systems integrators (SI) are small- to medium-sized businesses that help local businesses with their technology needs. Think local law firms, dentists, retailers, and corporate offices. They need technology, and these local partners are typically certified by the big technology manufacturers (e.g., Cisco, Microsoft, Google). You could work for them consulting, selling, and implementing. You should focus here because these companies will be the closeting thing to a tech company, in a non-urban area.

Technical recruiting firm—with the growing demand of technical talent, technical recruiting is needed everywhere. The purpose of firms like this is to reach out and find folks in the IT and developer gateways to help source and place them at companies that are hiring for talent. This is a great way to get experience in the tech recruiting space.

Transferable skills—Marketing, Finance, People, and the other gateways are needed in any business. A strategy could be to get experience in a gateway you are interested and find a way to transfer in.

Determining Your Where

If you are willing to move, go to an urban area. Move to your desired part of the world, but get close to the action. The bigger the city, the higher chance you'll find high tech. Just keep in mind that most exits by technology startups are happening from companies in Silicon Valley.

Here's the secret I've found with folks who don't want to live in the Bay Area forever. You come to the center of technology (or break in where you are. Remember my first tech gateway job was in the metro of North Carolina), build experience, make contacts, and later you are better positioned to set up a situation where you can work remotely or you've built the skills to work at a technology company in the area of your choice!

If you are willing to relocate, here are a few questions to help you get clear on a potential new location:

- » Is there a big city you are willing to move to?
- » If so, great. Let's focus on building your target list in that urban area.
- » Have you picked a gateway role you want to focus on? If not, go back to part 2 and pick one to focus on.
- » If applicable: can you weave in any location-related changes into your MAP as well to?

Remember—this is *not* a life-or-death decision. It's a stepping stone to the many doors that the tech industry has to offer.

GOLDEN NUGGET

The biggest takeaway from this chapter is to move to a big city. You'll put yourself in the best place to launch your career.

Note on a quick LinkedIn change—in the previous chapter, I advised you to change your LinkedIn headline to "Aspiring [gateway title]." Now I recommend you also include your desired location on your profile.

Congrats! Let's recap: from earlier chapters you already nailed down your TALENT profile and made steps towards your TIMING by using the MAP. In this chapter you've narrowed down your TERRITORY by pinpointing your where. Up next—building your TECHNOLOGY target list.

CHAPTER 16

BUILDING YOUR TARGET ACCOUNT LIST

Now that you've determined your particular gateway into tech, where you want to work, crafted an incredible resume, and a spot-on LinkedIn profile, it's time for you to do some thinking and research to build a list of the tech companies, small or large, that you are interested in working for. We're calling this your target account list.

For those who just want to get into tech and would rather get started building a list and attempting to break in FAST, this is what the first part of this chapter will consist of.

Here's the simplified plan for building your target list:

- » PULL ACCOUNTS searching for tech gateway openings at companies in the location(s) you have chosen.
- » PRIORITIZE your targeted openings into two buckets: your "A" bucket, which consists of the companies you are most excited about that you'll spend more time pursuing, and your "B" bucket, which are openings that you'll apply to but only focus on if you don't get traction with those in the A bucket.

STEP #1a: PULL ACCOUNTS: *Willing to Relocate to a Big City*

If you ARE willing to relocate. You'll likely get a lot more opportunities to work with. Here's how you'll PULL:

1. Create accounts on Angels List and LinkedIn (if you haven't already).
2. Search gateway jobs on Angels List, LinkedIn, Glassdoor, and other tech job sites like Indeed.
3. Search your gateway role in your desired urban city, and start building a list of companies with the roles you are looking for.

AND/OR

1. Google "Fortune best tech places to work for [insert end desired city]."
2. On that list locate the largest companies and then search their career portals to see if they offer your chosen gateway role in the city of your choice.

3. **Repeat these steps until you have a list of at least 20 companies with the gateway role you desire.**

Here's what your list should start looking like at this point (It'll save you time to save the URL of the career page):

Name of company	Gatway Role
Stripe	Account Associate
Docker	BDR
CloudFlare	BDR
DropBox	Recruiting/People
OKTA (OKTA)	SDR
Github	SDR
Anaplan	Sales
Twilio (TWLO)	BDR

STEP #1b: PULL ACCOUNTS: *Not Willing to Relocate to a Big City*

If you are NOT willing to relocate here is how you'll PULL to find relevant companies:

1. Google the phrase "Fortune best tech places to work for [current year e.g. 2018]."
2. Read about various tech companies, and then **choose a few** that you could see yourself working for.
3. Let's say you pick Salesforce.com, Cisco, Microsoft, Google (for example)
4. Google "how to find a Salesforce partner [or integrator] in [your current city]" or use the tech company's partnership locator (on their webpage) to find partners near you.
5. Search their career portals to see if they offer your chosen gateway role in the city of your choice.
6. Expand/change locations or select another tech company and repeat the previous steps until you have found at least **20 companies** with the gateway role/s you desire.

GOLDEN NUGGET

The idea is that if you are don't want to relocate, then there is technology being implemented everywhere in the world. For a big company, like Salesforce, Microsoft, or Cisco, they won't spend resources hiring people in more rural locations. They'll empower and work with partners and systems integrators to provide them training and certification, so they can scale their reach.

STEP 2: PRIORITIZE

Divide your 20 targeted companies into two buckets: your "A" bucket that hosts the targeted companies you most want to spend time pursuing. The "B" bucket hosts companies that fit the bill but that you aren't as enthusiastic about. If you are brand new to interviewing and reaching out to people COLD. It may worth practicing the techniques in this book on your B accounts. Why? Your first emails, calls, interviews are going to be rough. **Pro Tip: It's worth pairing up with a peer or mentor to practice and role play different scenarios. Find out more by going to www.irwinki.com.** When you feel more pass your peer & mentors satisfaction. Move up the list through your "A" target accounts.

Here's what your list should start looking like at this point:

A	CloudFlare	BDR
A	Github	SDR
A	OKTA (OKTA)	SDR
A	Twilio (TWLO)	BDR
A	Stripe	Account Associate
A	SignalFX	https://jobs.lever.co/signalfx/3
B	Quantcast	Recruiting
B	Weebly	
C	Wind River System	SDR Intern (Alameda)

* * *

THE IRWIN'S SILICON VALLEY

STRATEGIC ACCOUNT BUILD:

To get even more strategic and targeted in building your list, let's look at additional ways you can go about identifying companies that you could work for. To be totally transparent, this level of account building is really for urban areas that are known for having a VERY successful technology hub. If your location focused in Silicon Valley, I HIGHLY suggest using this strategic account build.

What I'm proposing here is very involved, meaning it will take a lot of time, persistence, and digging on your part, but the end result should be worth it.

Take your account list you built above. and do further research about the companies. You will research each company's:

1. FOUNDERS and your boss
2. VCs
3. STAGE and SIZE

These three factors could greatly influence the likelihood of your getting hired, the quality of your work experience at a place, and whether or not a place would even be the right fit for you. Determining these three factors for a company will provide you with the most informed decision on whether you want to pursue working at a certain company and how intensely you want to pursue it.

RESEARCH IS WHAT
I'M DOING WHEN I DON'T KNOW
WHAT I'M DOING.

— WERNHER VON BRAUN

Factor 1: Founders and Your Boss

Let's start with the importance of founders and your future boss. To show you just how crucial the founder-boss factor is, let's read what two experts have to say. When people ask Jason Calacanis, angel investor and author of *Angel*, "How do you pick billion-dollar companies to invest in?" Calacanis responds, "You don't pick billion-dollar companies (Jason has invested in startups that became billion-dollar companies: Uber, Thumbtack, Robinhood, DataStax, and Desktop Metal). You pick billion-dollar founders." Dr. William Raduchel, formerly a professor of economics at Harvard and an assistant dean at Harvard and Radcliffe, and currently an independent director, angel investor, and strategic advisor, advises, "Don't pick a job. Pick a boss. Your first boss is the biggest factor in your career success. A boss who doesn't trust you won't give you opportunities to grow."

Founder factor: in the startup world there are a lot of first-time founders, which means you can't do research on them to find anything about their prior successes. However, if you do research and learn a founder had prior success leading prior startups, then this will definitely increase the odds that they know how to build a culture of success and recruit and retain the right leaders and teams.

If you happen to apply to a very-early stage startup. Say <50 employees... there is a chance you are interviewed by one of the founders. If you made it to the final stage. They'll vet you for culture fit & in some cases your capability to execute the job. It's worth noting that one of the key responsibilities of the founder(s) is to OWN culture. If you have this opportunity, don't miss the chance to clarify their vision, technical chops, capability to lead, and questions that switch the tables and have the founder sell you on why you should join his company! Follow your gut and if you have multiple interviews at the same time...I'd take notes on what you learn from the different founder(s) to help you make this decision.

DON'T PICK A JOB. PICK A BOSS. YOUR FIRST BOSS IS THE BIGGEST FACTOR IN YOUR CAREER SUCCESS. A BOSS WHO DOESN'T TRUST YOU WON'T GIVE YOU OPPORTUNITIES TO GROW."

– WILLIAMRADUCHEL

Boss factor: Next you need to focus on the hiring manager, which is the person who would be your boss. If you are lucky, the right boss can become a mentor who can groom you, help you grow, and maximize your potential. They'll trust you and give you responsibilities that will challenge you. They'll also be able to lead from the front vs. command from behind like a dictator and throw you under the bus to save themselves. This is why getting know the boss during the interview process is key in ensuring you are putting yourself in a position where you'll grow regardless of how mundane the gateway job is.

Factor 2: VCs and Smart Money

Angels' and VCs' sole purpose is to invest in entrepreneurs' ideas to provide a return for their shareholders. They know most of their portfolio companies will not reap any return; however, if one does, then the return could very well be massive. Angels and VCs give entrepreneurs the resources, such as money, equipment, office space, etc., to run like a full business and grow quickly. These types of investors empower new ideas to occur.

The best Angels and VCs won't just give resources to any entrepreneur or startup. These investors do a lot of research. They carefully screen entrepreneurs who are looking for funding. They evaluate entrepreneurs in terms of their vision, technical chops, and ability to recruit and build a team to execute their vision. Last but not least, angels and VCs consider whether the market and timing are right for the particular product, technology, or service the entrepreneur/startup is offering. These investors screen hundreds of entrepreneurs before deciding who they'll back.

If you look at the graph below, from a Techcrunch.com article (2011), you'll see how paramount VCs have been in the fruition of successful technology companies. For example, during the PC Boom (1983–87) 13% of VCs accounted for 44% of the IPO value. During the Internet Boom (1997–2001) 4% of venture capital investment firms accounted for a whopping 66% of the value. What this shows then is that the top VCs have accounted for most of the biggest exits. If a company in your target list has a top VC, then that's a very promising sign.

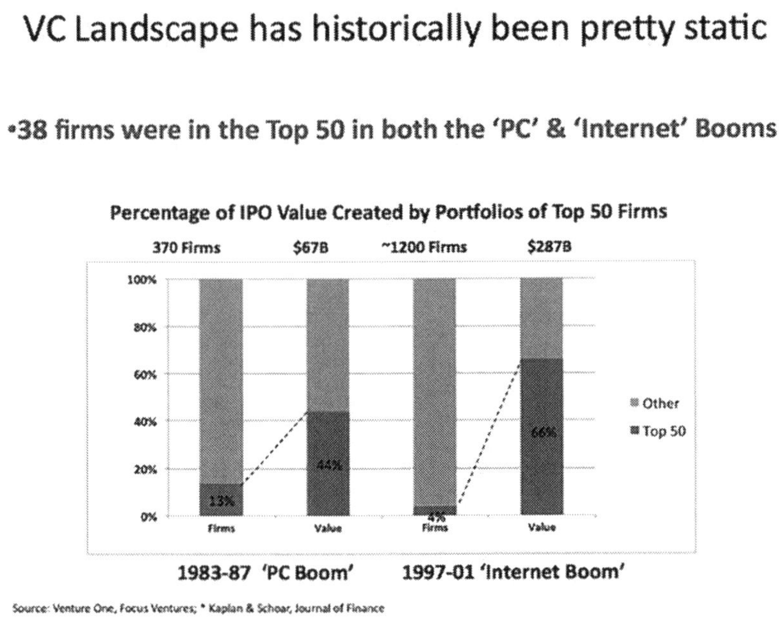

In summary, by determining which startups the best VCs and angel investors are backing, you are increasing your odds of finding startups that may exit. These startups are the ones that could be worth your time pursuing, so put them on your target list.

Here are the steps for determining the startups that the top VCs and angel investors are backing:

1. Google "top venture capital firms in [your desired city]"
2. For each firm, go to its website and locate its portfolio of startup companies it has funded
3. Go to each startup's careers page and search for your gateway position. If it's there ...
4. Add this to your target list.

Bonus: The Lazy Person's Way to Build a Good Target List

Read and review Business Insider's annual "Top (Enterprise) Startups to Bet Your Career On." It's evident that Business Insider has a full-time staff doing the due diligence and research for their annual article.

BI's "Top 25 Startups to be your career on in 2013" (12 out of 25 had exits)

- OKTA (IPO 2017)
- Cloudera (IPO 2017)
- Box (IPO 2015)
- Pure Storage (IPO 2015)
- Atlassian (IPO 2015)
- Coupa (IPO 2014)
- FireEye (IPO 2013)
- GitHub ($7.5B acquisition)
- Dyn ($600M acquisition)
- Appcelerator (acquired)
- Lithium (acquired)
- Pertino (acquired)

BI's "Top 21 Enterprise Startups To Bet Your Career On" 2014 (10 out of 21 had exits)

- Zuora (IPO 2018)
- MongoDB (IPO 2017)
- Nutanix (IPO 2016)
- Box (IPO 2015)
- Atlassian (IPO 2015)
- New Relic (IPO 2014)
- GitHub ($7.5B acquisition)
- AppDynamics ($3.7B)
- SimpliVity ($650M acquisition)
- Apigee ($625M acquisition)

In both 2013 and 2014 Business Insider articles almost HALF of the companies on their list had exits (Sold or IPO'ed)!!! In a world where more more than 90% of startups fail. Leveraging BI's annual article when making your next career move can put you in a great spot. Especially when you are new and not sure what makes a good startup.

Factor 3: Stage and Size

Just as your location greatly impacts the tech companies available for you to work at, there's another important factor that can impact whether or not you want to include a company on your target list: its stage of development. There are important pros and cons to consider when deciding to work at "newborn" company in its seed- or early-stage—e.g., ABRA—versus working at a mid-late-stage startup—e.g., Slack—versus working at a large and established tech giant—e.g., Amazon. So that you can make sure your target list includes companies that you truly would enjoy working at, let's review the pros and cons of working at tech companies at these different stages of development.

Seed and Early-Stage Companies

Remember the scene in *The Social Network* when Mark Zuckerberg first created his company with Eduardo Saverin and his roommate? In an instant, he created a 50:30:10 share model. In this scene, Facebook was in its seed stage. Everyone worked for free, and the company was running off of Eduardo's bar mitzvah money.

A friend of mine, I'll call him Aaron, was the first hire at Airbnb. He found the job posting on Craigslist. He was paid hourly to cold call people on Craigslist. He had to convince those people to move their short-term rental posts from Craigslist over to Airbnb. After a few months, he quit. To this day he regrets that decision, but at the time, when Airbnb was in this early stage, it was hard to see its future. Aaron was living in a very expensive city and getting paid only around $30k a year to do tough work. It just didn't seem worth it. However, if he had stayed, he would be extremely wealthy now.

Aaron's story illustrates what it's like to work at a company in its early stage. Particularly when a company is in its seed stage, it's running on a bootstrap. Everything is scrappy. There are likely no perks, very long hours, and a lot of grunt work. Chances are if you land in it, you are directly helping the founders doing executive assistant type of work. You could be their first sales hire (like Aaron was), and it likely won't be very glamorous (as was true for Aaron). In order to attract new talent, the startup may offer an equity package that sounds amazing, even 1–20% of the company if you are among the first few, but this equity won't amount to much if the company doesn't make it. Also, your wages will likely be low.

There are fewer gateway jobs available at these seed- and early-stages because these companies are in the process of proving their value and trying to get traction. The demand hasn't yet been proven. Most startups at this stage are in survival mode, and depending on the company, it is all hands on deck. You may be pulling 60- to 100-hour weeks. It can be a grind, but it can also be extremely rewarding, particularly if the company goes on to blossom.

Takeaways for Seed- and Early-Stage Startups

» Generally, pre-Seed, Seed, and Series A are early-stage.
» They are likely more flexible and willing to take chances hiring someone with little to zero experience
» Expect compensation to be below-average.
» Definitely expect a work hard/survival culture
» You learn a lot very fast.

- You need to believe in the vision of the founders.
- It can be financially rewarding, but it's like playing startup lottery with your time.
- This is the stage with the highest chance of failing (~70%).

Mid-Late-stage Companies

Mid-and late-stage startups that are in high-growth mode have the best odds of providing employees with personal and financial growth. They are likely to be venture funded (though some have been able to grow fast due to market demand). It's likely the founders have a clear vision of where they want to take the startup. They'll use that money to work fearlessly and grow! Everything seems to move fast at this stage. In board meetings the founders will have hiring goals alongside their revenue/user numbers. They must hire more people in order to continue the fast growth, which means they're looking for dedicated, hardworking people to join them.

You're going to find more of these types of companies in startup-heavy locations like San Francisco, New York, and Boston. Some examples of startups at this stage include Docker, Stripe, Mixpanel, and Slack, all of which have HQs in San Francisco.

If you want to work at a startup, I've found that mid- and late-stage startups make a great match for new folks trying to break in. Why? They are growing at a fast pace, but their brand name isn't as big as, say, Google or Facebook. Therefore, they are willing to take more risks in non-traditional candidates. The equity upside is still there for you, and the venture funding allows them to compensate you fairly as well. You don't have to eat ramen and use your savings (most of the time) to work there.

In return, you are expected to perform like an athlete. You are on a team, you have a specific set of duties, and the team expects high performance. If the company continues to grow, you'll get to see what it's like growing from 50 to 200 employees and 200 to 1,000 employees and then to 5,000. Each stage comes with new experience, and you get to watch and learn from the inside.

With fast growth comes great career opportunities. If you are a leader and you know how to build internal champions, at a growing mid- and late-stage startup you have the opportunity to get promoted much faster than at big companies. Remember the example in chapter 4 The Marketing Gateway about the marketing intern at Uber who went from intern to executive in a few years? It may have taken decades at a large company for her to reach that same position.

If the startup is purchased or exits via the public markets, you'll benefit from the upside as well as receive the money you made via your salary. It's a win-win. This gives you the foundation to grow into your dream career (and possibly gets you closer to financial freedom)!

Takeaways for Mid-stage & Late-stage Startups

» Generally Series B and C are mid-stage.

» Generally, Series D or later are late-stage.

» Because they are well funded, they can act and pay like bigger companies.

» As they mature, they'll grow and have strong leaders for you to learn from.

» Participating in their expansion and planting of the company's flag across the country and world can be exhilarating.

- There's less risk than with seed- and early-stage companies, and the upside can also be rewarding.
- If you love building and growing with a team, this is for you.
- If the startup doesn't grow or continue to innovate, it will die. It might happen a bit slowly over time.

Large Tech Companies

Large tech companies (I call these "enterprise" companies) include Microsoft, Amazon, and Apple. Then you have the leaders like Google and Facebook. Also, there are newly IPO'd large tech companies like Docusign, Dropbox and Snapchat.

I've worked at two such enterprise tech companies: BMC Software and then at Cisco through an acquisition. Both offered great benefits: 401(k) matching, employee stock purchase plans, and competitive pay. In comparison to seed-, early-, and mid-stage tech companies, the big boys certainly offer more job security and a better work-life balance, as well as structured training and processes. And you get to see how a very large company is managed. You'll see that they are the ones doing the acquiring versus being acquired.

Another benefit of working at an enterprise company is being able to include its "brand" on your resume. Every company has a brand associated with it. For example, a software engineer at Netflix carries more weight on their resume than one working for an unknown startup. I know recruiters who purposefully hunt for talent at enterprise companies. A brand name on your resume can enhance your overall career options.

Takeaways for Large "Enterprise" Tech Companies

- » Because they are established, you enjoy greater job security and much lower company risk.
- » There are longer promotion cycles.
- » On average, the pay is greater than at early- to mid-stage startups, and they offer excellent benefits.
- » They offer established training and processes (as opposed to you having to figure this out yourself).
- » Their brand name on your resume and LinkedIn profile enhances your overall career options.

From Readily Apparent to Mysterious

Enterprise Company & Late-Stage Startups = Readily Apparent

Determining if a tech company is at the enterprise level is straightforward. For one thing, they are publicly listed on the stock exchange. If it's a startup then they areon their Series D or greater in funding.

Mid-Stage Startup = Some Digging Required

These companies are usually in their series B-C phases. They usually have double or even triple digit year over year growth.

Very Promising Seed- to Early-Stage Startup = Hire Out a Backhoe, Lots of Digging!

It's easy to determine if a startup is in its seed- or early-stage: read my description of a startup at this stage and see if the startup you are doing research on fits the criteria: small band of employees, maybe angel-funded, bootstrapped, or just received their first angle or venture funding.

The crucial question for a startup in its seed- or early-stage then is: will it continue to grow? Will it break through and eventually exit? If it seems likely, then you'll want to include it on your target list as a place to pursue working at. But the reality is that most seed- and early-stage startups will not make it.

Solving the mystery—from my description of the pros and cons of working at a seed- or early-stage startup, perhaps you find yourself curious or even highly determined to work at one. If so, use the three lines of investigation I've presented to solve the mystery of how likely it is a particular startup is going to grow, succeed, exit, and IPO. After your investigation, if you come across a startup that you find promising, you'll want to include it on your target list.

Again PRIORITIZE

As I recommended at the end of the explanation of "The Quick Build," for this highly researched list, once you've built it, you'll want to prioritize it. Divide your 20 targeted companies into two buckets: your "A" bucket that hosts the targeted companies you most want to spend time pursuing. If you don't get traction in your "A" bucket, then move down the list through your "B" targets.

* * *

Identifying Influencers

After compiling your target account list, it's important to determine WHO are the (a) decision-maker(s) around hiring as well as at least (b) one employee already in the role you are applying for. This is key in increasing your chances of breaking in.

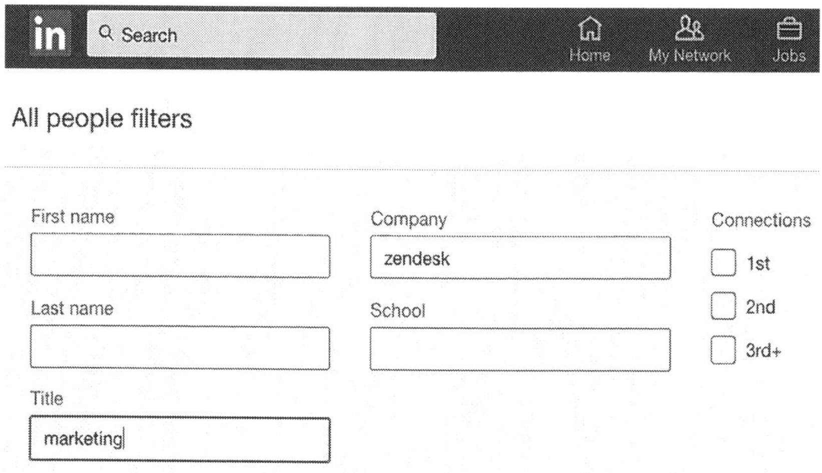

1. Pick one company at a time and go to LinkedIn.com
2. Use LinkedIn's search function and search title with [gateway] role with the company name.
   › For example, Marketing @Zendesk
3. Now search in LinkedIn and find at least one person within the company that you think would be your hiring manager (i.e., decision-maker or DM)
4. Now search in LinkedIn and find at least one person who might be your peer that you could contact for a

   could turn into a referral).

   Start building the list.

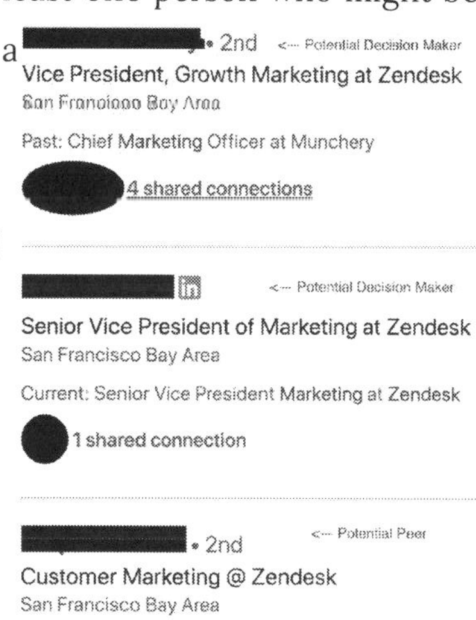

Here is what should show up if you did step #3:

Here is what should show up if you did step #4:

The idea is you want to understand who are the decision makers for the job you applying for and also build a relationship with your internal peer in hops that they may give you inside information or referral you for the open spot!

After compiling everything here is an example (names are made up) of what your target list will eventually look like:

Rank	Name of company	Gatway Role	Hiring manag	Peer (referral)	Phone
A	CloudFlare	BDR	Bill Capital	Tyler Rock	(415) 888-890
A	Github	SDR	Maria Case	Stevie Lau	(415) 888-893
A	OKTA (OKTA)	SDR	Bixby Park	Steve Pergolino	(415) 888-892
A	Twilio (TWLO)	BDR	Travis Fields	Frank Patterson	(415) 888-895
A	Stripe	Account Associate	Andrew Finkel	Jeremey Kutcher	(415) 888-888
A	SignalFX	https://jobs.lever.co/signalfx/3	Steve Middleton	Tamara Pechet	(650) 539-8650
B	Quantcast	Recruiting	Robert Erlick	Sara Gordan	(415) 888-889
B	Weebly	SMB sales	Catherine Marvin	Rustin Turay	(415) 888-891

Bringing It Together to Build Your List

Be sure to allow yourself an adequate number of days in your MAP when sequencing out this target building stage. Don't forget to remind yourself that the time and effort you devote to it will be worth it. This is your future career we're talking about, after all.

For now fill out the "Hiring Manager" and "Peer" fields along with your target account!

After you've put in the serious time and effort building your target list, it's time to initiate contact. The next chapter offers tips on how to go from random job seeker to sought-after applicant at the companies on your target list.

CHAPTER 17

BREAKING IN WITH REFERRALS

This is one of the most important chapters in landing your dream job. This chapter uncovers the insider's truth about why most people don't break in and the mindset and strategy needed to overcome it. Let's start with some facts and level set the perspective I'm coming from before going deeper.

FACTS

- Only 2% of online applicants make it to an interview. 2%!!!
- On average 40–50% of employees hired are referred into the company.
- Referral candidates are 3–4 times more likely to be hired than non-referral candidates.

» Companies with referral programs offer $1,000-$5,000 in cash incentives.
» High-growth companies use recruiters as needed to help find candidates FAST.

What does this all mean? I like to break down getting hired in one simple formula:

Problem + Decision-Maker = Opportunity

It's the same formula with breaking into a career. The employer has a **problem:** they have publicly posted, "We need HELP here!" If you can identify and connect with the **decision-maker**, then you put yourself in a great position to solve their problem, which for you is actually an **opportunity.**

The challenge is that most folks apply online and hope, but never put in the effort to identify and connect with the **decision-maker.** Most applicants never get a chance to sell to the direct **decision-maker.** Their resume is ignored, or they are immediately disqualified due to poor experience. You cannot be among this group.

I have seen time and time again the following scenario: there's a candidate with zero experience themselves, but they have friends in tech. With very little effort they manage to land an interview with **decision-makers** due to their connections (via those or friends). Connecting with **decision-makers** is vital.

In this chapter I'm going to share with you how you can break in and get noticed. You'll learn how to locate and connect with decision-makers so that you get the needed attention to land an interview. You'll read stories of people who succeeded at this, which should inspire you and motivate you to act. Here's the summary of what will be discussed:

» The power of brute force
» The power of informational interviews
» The power of referrals and understanding the game
» The power of selling directly to the decision-maker
» The power of external and internal recruiters
» The power of focusing on early startups

The Power of Brute Force

Shane just graduated from college and drove his beat-up car all the way from San Antonio, Texas, to San Francisco. He didn't have a place to stay, but he did have a Silicon Valley dream. When he arrived, he popped open his laptop and applied to every single sales-related opportunity on Indeed.com, Glassdoor.com, Linkedin, and the job portals on startups he was looking at. He had to make it happen. After selling his car when he arrived, he only had two months' worth of cash in the bank.

Three weeks later: Shane had seven offers! As we dissected his offers (see how I helped him negotiate in chapter 19) here's what he did during those three weeks to get these offers: he applied to over 300 positions, and he was juggling 4 to 5 phone interviews per day. From there he landed 28 first-round interviews. Of these 28, seven turned into offers.

If you look at Shane's results roughly a little under 10% of his applications resulted in interviews, which is much higher than the 2% of the national average. I suspect that his response rate was higher because A) he's a solid candidate and B) gateway jobs are in high demand. Remember, Gateway roles are not the most glorious jobs, and filling them presents a real need for hiring managers.

Here's the downside to Shane's method. After reviewing with him the seven companies that made offers, here's what we found. Two were way below the pay average, and only one had backing from a top 10 Silicon Valley firm. The crazy part was after Shane read the section on how to pick a well-positioned startup (in chapter 15) of this book, he DECLINED all the offers. And he did this with only one month of cash in the bank. He told me that he hadn't moved all the out to San Francisco to settle. What did he do next? Shane rebuilt his target list and hit the ground running again and landed his career at the startup he wanted!

The moral of the story: if you want to break into any gateway, you need to play the numbers. And if you are getting traction, as Shane was, it may be worth taking a more targeted approach. If you are not, it's worth getting in the game and building experience before moving on to other opportunities.

The Power of Informational Interviews

When you are breaking into a new industry, the "informational interview" offers the best way to learn what the career entails, what knowledge and skills you need to acquire, and how to best break in. An "informational interview" is not some formal thing. Typically it means sitting down to a coffee with an insider and picking their brain to learn all you can about the career.

For instance, I first learned about tech sales from a friend over breakfast. What is tech sales? What is a day in tech sales life like? Do I need tech experience? What do I need to do/have/be in order to be successful? Any advice on how to break in? Can you refer me or help nudge the recruiter to follow up with my application?

Because people who have friends and family have easy access to these insider informational interviews, it's no surprise that they often get called in for interviews.

One of the goals the gateways section (section 2) was to scale the tribal knowledge from my friends (i.e., tech insiders) so that you can get decades of informational insider knowledge and experience in just a few chapters.

Even with the plethora of info the gateway chapters provide, if you have a friend in tech, it's worth requesting time to pick their brain on the career you are trying to land. If not, the reason why in the previous chapter I recommended identifying both the decision-makers (DM) and future peers (folks at the companies already doing the role that you want) is to help you build a list of people to reach out to for informational interviews. The peer information interview is going to be the easiest to land. You can offer up coffee or take the person out to lunch, and they'll likely take you up on it. PLUS there is going to be more than one peer at each company that you can reach out to, meaning if one person says no, then you can simply ask someone else.

The one BIG piece of advice for reaching out is to reach out with a WARM tone and communicate how talking with you could be valuable to them.

IT'S NOT WHAT YOU KNOW,
BUT WHO YOU KNOW.

— UNKNOWN

By giving the person a reason or potential value, you help increase your chances of scoring an interview faster.

Here's a real-life example of one of my mentees:

Hi [redacted]
My name is Megan and I am looking to break into Sales Development. I have created an account list of companies to which I have applied; Salesforce landing at #1 of my desired places to work. As you are currently an SDR within salesforce I would love to chat with you and learn more about your position.
I understand that I am a stranger to you and this may seem out of the blue with little incentives on your end. However, I am aware that many companies have referral systems in place and if it is a mutual fit, I would be honored to have you refer me. With that being said, I would be very grateful if I could have ten minutes of your time, wether on the phone or in person over coffee (my treat!).
Please get back to me at your earliest convenience.
Sincerely,
Megan [redacted]

Here is the response:

Hey Megan,

Thank you so much for reaching out! I'm happy to hear you are considering Salesforce, it truly is an amazing company.

I would be more than happy to connect and give you a little insight on the role, along with a few tips and tricks. I would be free this week for a phone call, when works best for you?

The goal of the "ten-minute" chat with the tech peer is to get even more inside advice on what the ideal candidate looks like and to identify the gaps in your personal development that you can work on to become ideal. Also, it's possible that this tech peer or mentor can offer you a leg up to your dream job. How?

The Power of Referrals and Understanding the Game!

Remember:

» Fact #2: On average 40–50% of employees hired are referred to the company.

» Fact #3: Referral candidates are 3–4 times more likely to be hired than non-referral candidates.

» Fact #4: Companies with referral programs offer $1,000–$5,000 in cash incentives.

GOLDEN NUGGET

There's a HUGE factor in the technology that is typically unknown to outsiders. It's called "internal referral bonuses." Many tech companies will offer employees bonuses, ranging between $1,000 to $5,000, to incentivize them to proactively post job postings on their social media to entice the best people in their social networks to apply to the postings.

To give you an example, one of the original SDRs I hired has made more than $40,000 in referral bonuses. That's crazy! Not going to lie, I wish I had a stream of candidates to refer in.

The opportunity is to turn your informational interview into a potential referral or career mentorship. This is your best way to get your foot in the door! Get referred in.

Another way to connect with people in technology is to go to networking events or conferences in the field you are looking to break into. It's a great way to meet people who can connect you to their friends in the business. One of my first hires was referred in when my colleague's husband met my future hire at a networking event. My friend got a referral bonus, and I hired a great SDR who went on to be a great addition to my team. That's called a WIN-WIN deal!

Here's an example customized email, saying something along the lines of:

> *Hi [Name]! I noticed you have progressed in your career. I'm new to the Bay Area. I'm working on my resume, and before I apply for [insert gateway, e.g., SDR] position, I was wondering if I could pick your brain on what makes a successful candidate. Either way, I would love to learn about your experience. If you find it in your heart to help out, for your time I'm happy to send you a Starbucks gift card or take you out for coffee or lunch.*
>
> *Also, I heard most companies have an internal referral bonus. If you think I'm worth referring in, I would love to create a win-win scenario, especially if I were to be your [insert gateway, e.g., SDR] one day!*

It may take an additional email to other colleagues or following up with another set of emails, but with enough persistence, it will work. If you are able to convey why you'd add value and if you are reaching out to someone you might support or work under, then they may give you the time and advice you are looking for.

Please note: this coffee-date/casual chat approach to non-hiring managers will lengthen your tech gateway job search a few days because it requires extra effort and time (so include that extra time for this step on your MAP). However, this extra effort will pay off. It will help you get your resume on the top of the stack.

I've definitely noticed time and again that hiring managers at the places I've worked consider internal referrals with more warmth and openness. In fact, at the companies I've worked for, 40% of employees came from internal referrals. If you can master building out your internal referrals and network, you should be able to break into your first gateway interview fairly quickly. Either way, you'll eventually need to apply online!

The Power of External and Internal Recruiters

» Fact #5: High-growth companies use recruiters as needed to help find candidates FAST.

Another way to get "referred in" is through recruiters. There are two types of recruiters: external and internal. **External** recruiters sole job is to find **ideal candidates** and match them to their **corporate clients.** As a hiring manager, I have used external recruiters to jump start my sales team, and they did bring a lot of ideal candidates fast. For that extra speed and candidate pipeline, they charge the corporation 10–20% of the candidate's salary if they are successful. When I graduated, I found an external recruiter who lined up three interviews the very next week after my first call with him. The power of external recruiters is they typically have a list of clients andcan get you interviews veryfast!

BUT ...

External recruiters may work with companies that may not be so great. It's very important you do homework researching the company that they are lining up interviews for and really ask yourself whether you want to be "pulled" to your next opportunity or you want to "drive" towards a company you can see yourself building your next career with. Either way is fine. You just need to be aware of that. I recommend reaching out to external recruiters AFTER you have reached out to your target accounts as external recruiters are a great way to build your pipeline of interviews fast. They can be used to help you get extra interview practice and get multiple offers (see chapter 19).

Eventually, a tech company will invest in building out their internal recruiters. These are recruiters that are hired by the company whose sole purpose is to understand the hiring managers' and company's initiatives; then they hunt for their ideal candidates as well as screen resumes.

Internal recruiters can be your best friends or worst enemies. Internal recruiters are the first gatekeeper in getting in front of the interview process. If your resume does not fit their standard, you'll go into their NO pile.

BUT if they like you, they'll become your internal coach to give you advice and foresight into the process.

Takeaways: Recruiters are a great way to build interview pipeline. If you change your LinkedIn to "Aspiring [gateway position, e.g. SDR] as previously mentioned, you should get recruiters reaching out to talk to you. You need their support.

The Power of Selling Directly to the Decision-Maker

Remember only 2% of online applicants get an interview, so if you do not get a response from applying online, then there is always going straight to the decision-maker (DM).

The whole point of building internal referrals and getting interviews via recruiters is to get in front of the decision-maker (DM). Remember:

Problem + Decision-Maker = Opportunity

This is the least-taken route due to the amount of courage you need. It is much easier applying online and hoping. It is much easier getting an external recruiter to reach out to you. What isn't easy is reaching out to a peer for a referral, which we already discussed, or this route: selling directly to the DM, whether that be an internal recruiter or the hiring manager. It takes confidence. For those who are willing to do what it takes, it does work!

Remember the job posting is a sign that there exists a **problem.** There is an incredible amount of effort and money spent to find the right candidates. If you can get in front of the DM you've already increased your odds of success. The hardest part when starting out is getting in front of them.

Here is an example:

Kirsten's directly selling to the Decision-Maker

I received a resume from a candidate named Kirsten who was applying to be on my sales team. At initial glance, Kirsten's resume was not that impressive.

She had few sales experiences, and the red flag was that her last three jobs were in 4- to 6-month stints. She was, in my eyes, a job hopper, i.e., not yet mature enough to make a career decision and stick with it.

A week later, after I dismissed her resume, I got an email from my VP in which he forwarded the email Kirsten had sent him. In it, she introduced herself to my boss and explained why she was a great potential candidate. It went something like this:

> *Hi Mr. VP,*
>
> *I applied online for an SDR position last week and haven't heard back. I was a top performer at X Company and Q Company in Los Angeles. I moved to San Francisco this month, and I saw that OKTA recently raised a new round of funding. Congrats! Likely your team would like to invest that money into sales reps that will challenge and elevate them.*
>
> *Would you mind giving me 15 minutes of your time?*
>
> *Warm Regards, Kirsten*

When your boss sends you a candidate, it's very likely you are going to give the person a phone call, at the very least. That was true in this case. Even though I was initially underwhelmed by her resume, I initiated the interview process. Much to my surprise, in our initial phone call, I found out that Kirsten was smart, articulate, and ready to get serious about focusing on tech sales. She explained her "hopping" as due to moving to the Bay Area because of her boyfriend's career change.

After an extensive interview process, I ended up hiring Kirsten, and she's shown herself to be an invaluable member of our sales team.

The big takeaway: Kirsten would have never gotten the opportunity to explain her side of the story and to address my concern about her "hopping" if she hadn't reached out to my boss. Kirsten's strategic email to my boss got her foot in the door, so that she could be heard and eventually hired. It was a brilliant move on her part.

While, of course, every scenario and hiring manager are different, I'm sharing this reaching-out-directly-to-the-DM technique because it can work. This happens all the time, and not just in hiring. In tech sales I often reach out to a VP or C-level executive with a customized email, and they, in return, refer it down to their team to evaluate and report back.

How to cold email like a #1 SALES REP!

Step 1: Find the right players

You locate the VP, Director, or manager (these are all decision-makers) on LinkedIn.com. Every gateway has a direct manager and even multiple managers that is responsible for building out the team. You need to do some guesstimation on who you need to locate.

For example:

» Recruiting coordinator -> recruiting manager/director of talent, VP of people/HR
» SDR -> SDR manager, VP/director of sales
» Office coordinator -> VP of people/HR, people operations, CEO/founder

Step 2:Find the person's email

Find the company's email structure. You can Google: *[name] + email address.*

» You'll find that you'll get an email structure like *first_initial last_name@company.com*

» Services like Thrust.io, Hunter.IO, Clearbit, Sellhack, or Headreach can help you find almost any corporate email address format. All you need is the company domain (e.g., @google.com) and first and last names.

» Verify the email address, leveraging services like https://verify-email.org.

Step 3: Craft an IRRESISTIBLE email

Research the company and their Linkedin page to draft a compelling, targeted email or Linkedin message. Kirsten's email above is one example. The most important thing to remember is **make sure your "ask" clearly shows how it benefits the receiver!** Kirsten's email emphasized she was already a high achiever and that will benefit the VP. Every email is going to be different just make sure your email isn't just me me and me. Rather you'll benefit because of X, Y, Z reasons.

Step 4: Follow up and CLOSE!

Follow up two days later and continue following up until you get a response (yes or no). These executives get a lot of emails. If they ignore you. Don't take it personally. As long as you're being respectful, you can send 7 to 10 emails every 5 days without being annoying. If they do respond...respond FAST (ideally within 5-30 minutes)! This is your chance in making a killer first impression and landing the next step!

GOLDEN NUGGET

Making it easier for the person to follow up is key. One of the ways to do that is to proactively offer up at least 3 different days and times you are available for a call or meeting.

Step 5: There is always another person

If you get no response at all, reach out to another decision maker and continue to line up an information interview with a peer to help get referred in. Remember there are more managers than directors, more directors than VP's. Someone in the chain of command may be more responsive.

BONUS:

Send thoughtful gifts. Sending well researched or thoughtful gifts will leave an impression. One story I didn't shared was how I got my first tech interview. Initially my friend referred me and it was great...BUT the company went on a hiring freeze. I stayed in touch every month until holy crap the freeze was over! When I found out I immediately bought a bag of coffee and attached my resume with a note:

> *Hi Randy!,*
>
> *Thank you for considering me for the SDR role. I heard that your hiring freeze is over. Here is some coffee to help things thaw faster :) When can we pick up our last conversation? I've attached my resume for your convenience.*
>
> *- Warm Regards, Irwin*

It worked! Got an onsite interview the very next week and an job offer that has forever shaped my life forever the following week after.

REMEMBER: 6 degrees of separation

Have you heard that you are only are only 6 people away from ANYONE on this planet. This includes celebrities, politicians, and of course your hiring manager! There was a Facebook Application named "Six Degree" and had over 5.8M users before it was shut down. And it showed that there was 5.73 degrees of separation. What is crazy is that Facebook released their own study that in 2016 that the distance has decreased to 4.57 degrees!

In Chapter 14 I asked you to connect your contacts to LinkedIn this 4.57 degrees of separation is why!

A. LinkedIn 1st-degree: People you are directly connected
B. LinkedIn 2nd-degree: People who are connected your 1st-degree connections
C. LinkedIn 3rd-degree: People who are connected to your 2nd degree connections

The goal of the "Power of Referrals" section of this chapter is to share how you shift your zero-degree connections into 1st degree friend who can refer you in.

Now let's talk about how we can leverage your network to get you introduced your desired decision maker. Let's say I wanted to reach out to CEO of Zappos Tony Hsieh

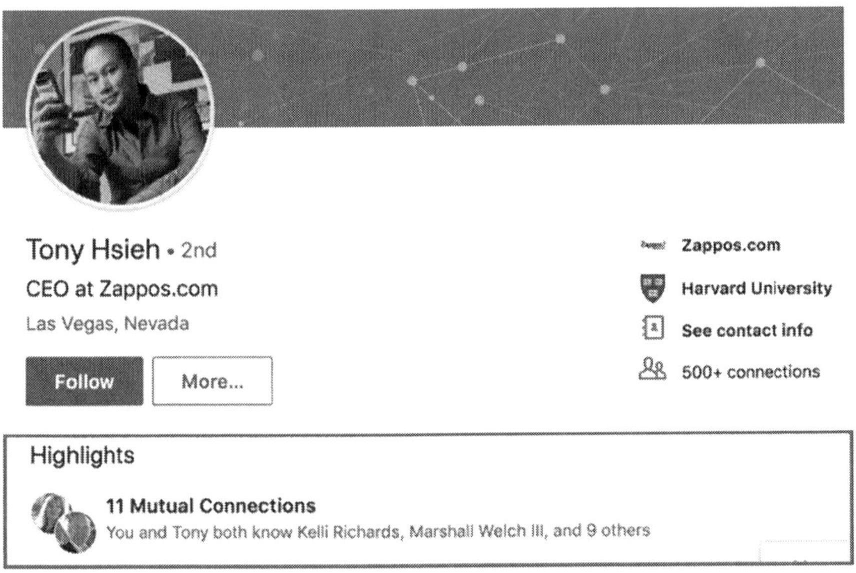

I have 11 contacts that I can ask and see if there is a mutually warm intro they can help make. If you don't have any mutual...no worries you are just starting out. Just go back to cold emailing.

Ask for the intro and make sure you have written majority of the email for him. I do this a lot in software sales when I'm ghost writing an email for my VP of Sales to send to one of my executives. Remember he or she is doing me a solid and making it EASY helps improve your chances that they may help you out! Here is a an example of a mutual email intro template:

> *Hey _____,*
>
> *As two people I have a ton of respect for, I'd like to introduce the two of you.*
>
> *Tony-Irwin is a buddy who I met at Disrupt SF a while back. He's been selling to IT guys his whole career* (BMC, OKTA, Cisco, etc) *and now he is helping newbies break into the tech scene.*
>
> *Given you have written a book on building a corporate culture (Delivering Happiness). I thought there was mutual synergies and wanted to connect you two authors.*
>
> *Warm Regards, Kelly*

After getting introduced. I should respond and explain my ASK as fast as possible and align it with why Tony would care. I hope this helps paint a good idea on how to ghost write a warm intro for whomever you are trying to get in front of.

The name of the game is getting as many interviews as fast as possible. These strategies will help get you a shot, so you can learn what is holding you back from making it happen.

Seeing It in Action + The Power of Focusing on Early Startups

I want to share with you a real-life example of how Tristan Walker broke into Twitter and Foursquare before they were huge. I first heard about Tristan's story on Master of Scale, the podcast made by Reid Hoffman, co-founder of LinkedIn and partner at Greylock.

Below is an excerpt from the podcast's May 2017 episode titled "The Beauty of a Bad Idea" which gives Tristan's example of breaking in by finding a referral as well as directly reaching out to the **decision-maker** and solving one of his **problems!** Notice too that there are three people in the interview. In addition to Reid Hoffman and Tristan Walker, there's David Hornik. You'll soon understand why David Hornick is included in this interview too!

Hoffman: That's Tristan Walker, the founder, and CEO of Walker and Company Brands. Their flagship brand is Bevel, a razor for men with curly facial hair. He started as a kid in the projects ... Tristan doesn't sound like your typical Silicon Valley geek. But I can't stress this enough—you don't have to wear a food-stained hoodie to succeed here. Neither do you have to play World of Warcraft or write code until sunrise. You only need to have a sense of curiosity—and Tristan was exceptionally curious. He wasn't just a student of business, he was a student of every technological shift taking place around him.

Walker: I was in accounting class at Stanford, and I remember MC Hammer was supposed to speak at Stanford. And there was this commotion going on—people wondering if he was actually going to do it. It was going around for five minutes, and I opened up Twitter, and I just asked MC Hammer, "Are you coming?"

Hoffman: Bear in mind this is 2008, back when Twitter had a relatively cozy community of 500,000 monthly users. Tristan was one of the more active members. As luck would have it, it was Hammertime on Twitter.

Walker: Thirty seconds later he replied back, and I turned around to my fellow classmates and said, "Yes, he's coming. See?" At that point, I realized how important Twitter's part in the innovations within communication were. And it was my first understanding of bad ideas being a good idea—because everyone else around the table was like, "Why are you on Twitter? What's the point of this thing? I don't care about what you're eating for breakfast." They just didn't get it. And that showed me that there was something there that I had to kind of dive into.

Hoffman: So we're already seeing the seeds of Tristan's success. You don't have to be a geek, but you do need to geek out over the new ideas. You want a level of enthusiasm that's slightly out of step with your peers. And Tristan was seriously geeking out over Twitter. He wasn't just an early user—he wanted to help build the company.

Walker: So I emailed 20 different folks who I knew were either one or two degrees separated from the company. The last person I emailed was David Hornik because he was a professor at Stanford and also a partner at August Capital. He replied back. He said, "Come in my office, let's talk." David asked me two questions.

David Hornik: Tristan, are you willing to work for free for the summer?

Walker: [Laughs] And I said, yeah.

Hoffman: I'm happy to introduce you to Twitter. I'm curious if you actually know what
Twitter does.

Walker: Yes. And at the time, they weren't thinking about the business model, and I was fascinated by what it could become.

Hornik: Send me your resume, and I'll send it to Ev.

Walker: Probably two days after that, Evan Williams [founder and former CEO of Twitter] sent me an email saying, "Come in for an interview." And a week after that, they gave me an internship at Twitter. I worked there for about six, seven months when there were 20 folks at the company.

Hoffman: You may think Tristan is lucky to get an email from Ev Williams, but notice the size of Twitter's workforce at the time—20 employees total. Tristan spotted the company's potential not just ahead of his classmates, but ahead of the market. He has a knack for spotting ideas right before blast-off. David Hornik noticed this too.

Hornik: Twitter was nothing when Tristan was excited about it. Twitter was a place that you told your buddies you were heading to the park. Twitter, at the time, was a service where you said, "I'm eating a delicious ice cream cone." It was entirely unclear what one did with Twitter, how one could make money at Twitter, how big Twitter would become.

Hoffman: Shortly after Tristan's Twitter internship ended, he started his next email campaign. He bombarded the founders of a fledgeling startup called Foursquare. And again, the CEO, Dennis Crowley, responded.

Walker: I emailed them eight times—the eighth time Dennis sent me an email—I'll never forget, this is verbatim—he said, "Tristan, you know what? I just may take you up on some of this. Are you ever in New York? Dens." That's it. I was in LA at the time, and I was sitting on the couch with my wife, and I said, "How should I reply to this guy?" Ten minutes later, I sent him an email and I said, "Actually, I was planning on being in New York tomorrow." I booked my flight that night, flew out the following morning, hung out with them for a week, and a month later I was running business development for the company.

*

One thing that wasn't mentioned was how very badly Tristan wanted to break into tech. He wanted to break in so much that he took it upon himself (while he was still a student) to cold call businesses to gather leads of customers who would be interested in advertising on the Foursquare platform *before* he was hired on at Foursquare! By doing this, he was proving to Foursquare that he could add value to them; he was already solving the **problem** (acquiring the customers) by getting in front of the **decision-makers.** An effective way to get any company's attention is to surprise them by solving one of their problems before you even work for them!

Another lesson learned in Tristan's story is the power of landing at a startup when it's still in its early development stage. If you are having a hard time breaking in and you've already made at least 10 attempts at reaching a DM, consider concentrating on a smaller early-stage startup. This will increase your chances of a response. Small startups don't get the overwhelming number of candidate applications that large established brand names get bombarded with.

Even when you contact the DM of a small startup, use the formula. Focusing on the **problems** and **decision-makers** at the startup. This is a recipe for success.

The Importance of Mindset

I have been mentoring Leah, who is looking to moving away from server work and into tech sales. First impression: she's a rockstar. Leah had great presence, and she knew how to upsell customers—heck, that's how we ended up working together! She impressed me.

I spent a tremendous amount of time educating her about tech sales (her chosen gateway) and tech recruiting, crafting her resume, enhancing and improving her Linkedin, and building her target list. In fact, all the steps in this book describe her path to breaking in.

When we got to the stage of reaching out to her target companies, this is where I realized the importance mindset. When Leah applied online, as expected, she got one rejection and the rest sent no responses. The result: she got discouraged and started missing our weekly meetings.

Then we got a break. I helped referred her to a tech friend of mine who was actively looking for SDR roles. Leah killed the phone interview, and that led to two additional onsite meetings. However, in midst of the interview process at this friend's place, Leah stopped pursuing her other targets. She stopped trying to reach out to the other companies on her list to locate peers/referrals or reach out directly to decision-makers.

It seemed Leah was banking on only one opportunity. She got a ton of messages (15–20!) on her LinkedIn from recruiters. I asked her to follow up with them and get the interviews lined up. She only reached out to 2 of the 15–20. Then her car broke down, then it was her work, her errands. This is when I realized that while not everyone may want work in sales, to land a tech job fast, you need a sales mindset!

Here is that "sales mindset" that I think is required to land a tech job in 90 days:

1. I need to email and request a coffee informational meeting with my "peers" in ALL my target accounts.
2. I need to write a targeted email to the decision-makers in my ALL target accounts.
3. I need to constantly be finding my "gaps" in my gateway skill set (excel skills, HR certification, toastmasters skills, etc) and taking a proactive stance on filling gaps with courses, books, or training.
4. I need to research networking meetups and attend at least one per week.
5. I need to respond to, close for next steps, and schedule calls with ALL the recruiters that message me on LinkedIn.
6. If I get NOs...I need to have the courage to ask why and learn from the feedback and grow.

Remember, there is more than one peer at each of your target companies. There are even 2–3 different decision-makers in each company. There are plenty of people to help you break in. If one says no or doesn't respond, you keep going. You email someone else. It requires playing to the fullest in order to achieve a breakthrough.

Looking back at my career, I see that it took five onsite interviews for me to land my first offer. **This is what it takes.**

Eric Schmidt, who was the CEO of Google and is now chairman of Alphabet, said that persistence is the single biggest predictor of future success ... and the second thing is curiosity: "The combination of persistence and curiosity is a very good predictor of employee success in a knowledge economy."

Without an unstoppable mindset, I believe most will give up and never break into the tech career of their dreams. It's important to discuss the perseverance needed to break in (see Jim's interview in chapter 10 where he tells how his resume never made it past HR and he hunted down the hiring manager). Technology is a huge growth industry and opportunity. It rewards the folks who want it the most. The biggest reason why people don't break in is a lack of perseverance; they aren't persistent enough to keep on and on and on, in the face of many obstacles, to achieve the break they're seeking.

The goal isn't to get a job. It's to take consistent action to land interviews and to keep growing in the process until you succeed. That's how you'll break in.

CHAPTER 18

SLAM-DUNKING THE INTERVIEW

I used to think that you needed credentials and fancy degrees or skills to land a tech job. After working in this industry, hiring, and helping others get started in their careers, I've learned that there are typically two questions every interviewer asks themselves:

1. Do I like (and will the team like) this person?
2. Could this person learn to do this job well?

That's it.

I've developed this simple formula for crushing your interview:

Likability+Capability = Hireability

The interview process includes interviews by management and peers. The purpose of the peer interview is to see if you get the team's approval that you'll be a good fit. And the peer interview carries as much weight as that by the management. I've seen candidates that didn't get positions because a peer didn't like them in the peer interview.

This section will help you focus on how to make a great first impression with management and peers. If you can work and master what I explain about likability and ability, it will help with your information interviews, networking, and how you approach your first phone and onsite interviews. You will crush all of it!

Likability

Here is a key section of Dale Carnegie's timeless book *How to Make Friends and Influence People*. The goal is to give you a few tips to further increase your likability by making a positive first impression and building value in your new relationships.

Here are Carnegie's "Six Ways to Get People to Like You" as given in *How to Make Friends and Influence People:*

1. Become genuinely interested in other people.
2. Smile.
3. Remember that a person's name is to that person the sweetest and most important sound in any language.
4. Be a good listener. Encourage others to talk about themselves.
5. Talk in terms of the other person's interests.
6. Make the other person feel important—and do it sincerely.

In the context of landing your first informal chat, phone interview, or onsite interview, applying these six steps can drastically improve your success average (as most of us were never taught how to get people to like us in less than 30 minutes, which is the average scheduled time for a chat over coffee or first phone screen).

Here are some of my suggestions based on the steps:
On step 1, "Become genuinely interested in other people": I've found it helpful to do research on the person I'm meeting and-or show real curiosity about their interests. I've found that people love talking about themselves, and more importantly, everyone has deep wisdom. Consequently, if you can stay authentically curiously engaged with someone, you might learn something new. It helps if you are practicing step 4, "Be a good listener." If you are able to do 80% listening and 20% asking better questions, you may be able to learn a person's deepest **needs**, which falls under step 5, "Talk in terms of the other person's interests."

One of the secrets I've found when building new relationships is to find out what a person **needs** and then see if I can help or if I can connect them to someone who can help—with no strings attached. I'm simply paying forward and adding value to people I meet in my circle. I've found that this elevates my respect and friendship as the people I'm connecting them to also have a need or service that they can help with. Having a **service mindset** and **helping others** will build your value fast!

On steps 2, 3, and 6: it's a great reminder to smile, remember another's name, and ensure the person feels important without any of this being fake. Being warm and adding **energy & enthusiasm** to your personality will help with improving your likability. I can't stress this enough. Doing

LinkedIn research on your interviewers can help create talking points that can help with rapport. The reason to dress well is to build your likability as well.

Likability is key.

Capability

Now let's talk how to increase your chances of showing your capability and at the same time avoid any mistakes which could undermine you. To best demonstrate how **capability** can play out in an interview, here's a description of an interview situation I once experienced.

The candidate, Mitch, had been through two phone calls, two on-site interviews, and one presentation. Mitch passed the capability tests with flying colors. I could tell he has invested in public speaking skills and he has been reading books on sales (adding to his skill set). This candidate was now in the final interview with Gustaf, the startup co-founder, and final decision-maker. Gustaf would give the yay or nay on this candidate, who would be my fifteenth SDR hire.

When Gustaf walked out of the conference room, he informed me, "We can't hire him."

Inside I was yelling, "NOOOOO!"

"Why not?" I asked calmly.

"He doesn't show much interest in becoming a sales rep. He eventually wants to go back and get his MBA. I don't think he knows what he really wants. We can't afford to train reps who work for one or two years and then get up and leave. I'm sorry. It's a no."

"I'm sure I qualified the heck out of him," I was thinking in my head.

After I walked the candidate out, I decided that I had to resolve my confusion. I returned to Gustaf's desk. "Gustaf," I said, "I'm scratching my head. How did you know Mitch wanted to get his MBA? When the other five folks and I interviewed him, we didn't catch any of that. I'd prefer to get this out earlier in the hiring cycle, so we don't waste so much time in the future."

Gustaf explained, "When I'm done with my interview questions, I purposefully shift gears and become more casual and open. Then I present something like this to a candidate, 'Great! Sounds like you're an awesome fit. Let's pretend for a second that we meet up in five years from now in a bar. I say to you, 'You look amazing! It's been a while. How's it going?' You say, 'Life is great!' I ask, 'Why is life great? What's happened since working together?'"

"Okay. What did he say?" I asked Gustaf.

Gustaf told me, "Your candidate said, 'It's great because I just wrapped up getting an MBA.' That tells me his deep down plan and desire is to get an MBA versus the software sales route. We're a startup, and we need to focus on hiring reps who are here for the long haul."

Wow. Gustaf was right.

LOSERS REACT.
WINNERS ANTICIPATE.

—TONY ROBBINS

This story is a great example of not aligning to **the needs** of the final decision-maker. Mitch got the entire hiring team to like him and proved to everyone that he could be successful in the role. Which is why he made it to the meeting with Gustav. What happened was he overshot his skis by oversharing what he wanted to in the near term that didn't match up with the long term vision of the founder. His likeability dropped in an instant. This is a good lesson the importance of not oversharing and aligning to the hiring managers needs.

Making It or Breaking It

After doing hundreds of interviews and learning from very smart hiring managers, like Gustaf and my mentors, I've seen both the common pitfalls and the successes of candidates. The rest of this chapter is going cover the ins and outs of likability and capability, including red flags, mistakes, and best practices in preparing and nailing your interview, so you can speed up your time to getting offers! If you want to take an inside look at what goes on after the interviewee leaves the room, I'd suggest paying close attention.

High Interest (Likability)

Be sure to mention activities that show you are very invested in the particular gateway you are looking to enter and the specific company too. For me, because I was interested in sales and software, I may mention one of the following in my interviews:

» I read *The Challenger Sale.*

» I attended Dreamforce (a tech conference).

» I attended a social selling online course.

» I'm a devoted listener to Reid Hoffman's "Master's of Scale" podcast.

A Focus on Strengths (Capability)

Your resume is likely not going to have a lot of meat, so focus on your strengths in the interview. Be present for your interviewer and show them why you'd be a great hire. Don't bring up where you are lacking (i.e., any lack of experience or low GPA). Focus as much as you can on where and how you are strong and who you'll become for them.

Conviction (Capability)

I had reps who made it to the final interview with our co-founder and VP of sales, but then they made one tiny mistake that caused them the opportunity. They didn't convey with conviction that sales (or your chosen gateway) were where they really wanted to be. When we were building out the team, we wanted to invest in candidates who could be the future leaders of the company. We wanted that longevity in their vision.

Loyalty: Now and in the Future (Likability)

Do not say anything that will show a lack of loyalty in the future! Think of the story with Mitch: you want to get an MBA, travel the world, or pursue XYZ passion in a few years? A candidate who wants to get an MBA suggests they aren't worth the years of training a company puts in new hires. It suggests your heart isn't even in the company because your dreams are elsewhere. Companies want employees to be with them for the long term, so express that as your vision in your interview.

Your Whys (Capability)

Here are the three most common whys a hiring manager is looking to find out in your interview:

1. Why do you want to do this job?
2. Why do you want to work here?
3. Why are you better than the team I have? How can you elevate and challenge my existing team?

If you can nail the three whys, it'll set you up nicely. Most people don't take the time to finetune their answers. Great leaders and experienced interviewers will always try to poke holes and test how you answer these questions. You'll need to back-up all your answers.

Task 1: go to the camera on your smartphone and record yourself in a video answering these three whys.

One of the best ways to improve your interview skills is to practice and record yourself pitching, watch yourself, and make notes on what you like and don't like. You also need to repeat the exercise until you convince yourself you will be a great hire. Please don't wing it. Trust me.

GOLDEN NUGGET

It may be uncomfortable. But by practicing with a friend, mentor, or family member...having practice selling yourself will help smooth out the rough edges most have when they first start out.

Finding your why is the most crucial piece of advice I can give for the interview process, as it will drive your conviction, confidence, and how you show up in the interview.

Task 2: after you figure out your whys, it's essential to start visualizing your desired outcome. This ties in because it will help craft your story and help the hiring manager see where you want to take your career. When hiring, there is a risk for the hiring manager on whether or not you are the right person. Come to the table with clear answers to these three whys and paint a convincing picture as to where you want to go. Your inner drive and thought-out vision will reduce the manager's perceived risk of hiring you. To the hiring manager, they see why you want to work at their company and where you want to go. If where you want to go fits within the company's landscape, it addresses the manager's concern about whether you are there for the long haul and whether you will help build and grow with the company.

I want you to complete the following exercise:

Write out your desired outcome at the end of your first year at the company.

Here are a few examples (Insert your desired outcome below):

1. I want to learn the skill of .... by .... (insert date).
2. I want to work for the
3. I want to pay off my student loans and credit card debt by ....

Make sure you take time and write out YOUR one-year goals at your first tech company, so you can mention these in your interviews. Goals without deadlines are dreams. So make sure you incorporate a deadline.

When Looking for the Ideal Candidate

There are aspects of the interview that I always take note of. For example, I pay attention to whether the candidate shows up on time. If they don't, I note whether they communicated any change in time or expectations. In the interview I task them with delivering a presentation or task involving the technology I'm hiring them for. This task has a tight deadline. I want to see how they'll perform against other candidates with the same time constraint and prep materials. I'll dig into their resume and try to find a situation in their past when they encountered roadblocks and ask them to speak about it.

Even still, I have a compass of five qualities that are most important in regard to the people I hire. I've used the following compass for all my hires. It has been a great guide, and the results of nine out of the ten SDRs I've hired have turned out to be fantastic. Here are those five essential qualities:

» Integrity
» Coachability
» Fire in the belly (inner drive)
» Intelligence
» Humility

Why Integrity Is #1 (Capability)

My definition of integrity is being one with your word. I'm looking for those who give their word and do everything in their power to keep it. This is not to be confused with morality. This was a huge distinction for me in college. A person with integrity makes a commitment and holds true to it. If something changes, then they will proactively communicate the changes in order to keep their word.

Without integrity my team can't perform at its highest possible level. Period. Imagine a bicycle wheel and one of the spokes is removed. In the Tour De France if you put that wheel on a professional's bike, it's a guarantee that over the course of the race, the same cyclist with the fastest times have complete spokes on their wheels versus a wheel missing a spoke because that wheel, even with just one spoke missing, is out of alignment; it doesn't move truly. That's the simplest way to explain it. Low integrity = lower the results. If I can weed out the reps who already have trouble keeping their word, my team is going to be on a different playing field.

A Tale of Integrity: Connor from Seattle

One of my top hires was Connor from Seattle. In the interview, he made a lasting impression on me when he shared his story about his quitting his job at Enterprise Rent-a-Car to travel Australia with his friends. He had just wrapped up the trip and moved from Seattle to San Francisco.

When I asked him why he quit Enterprise to travel, Connor shared that he and his buddies always said, "Someday we're going to do a backpacking trip through Australia" and "Someday I want to get into software." What made Connor stand out was that he was a guy who took his vision and made it happen! Without a backup job, without fear and excuses, he just did it! I respected his actions and knew he had power behind his word. He had integrity.

You can generate noticeable power from giving and keeping your word. It also shows integrity to proactively communicate when things change (i.e., you are running five minutes late) or when you've made an error. I challenge you to become that person who can voice a big commitment and find a way to make it happen. Just think of Connor and how he took action to get to where he wanted to be.

Coachability (Capability)

Ever since day one I've been expected to learn and grow. Every week we had product training. I learned how great and mediocre employees manage their businesses, calls, and time. I learned that everyone has something they are great at, and it was my job to find that golden knowledge to accelerate my skills. When I started progressing through my career as an SDR, SDR manager, inside sales rep, and field rep, I saw that the best have never stopped learning. They are always growing and challenging themselves. What I call this quality is coachability.

Coachability is so important in the technology business. Software attracts very smart people, and it's a great learning opportunity if you can seize it. Most people who start in technology sales have no prior tech experience. It's up to the managers and team to help empower new hires. With a lot of learning come a lot of mistakes and feedback. The best applicants are the ones who seek out coaching and embrace it as a tool to transform themselves into a better rep every single day. The worst applicants are the ones who see coaching as getting attacked and criticized.

One of the things that I fell in love with regarding technology is the amount of personal growth that comes with it.

So how do I personally determine the level of someone's coachability in the interview process? Easy! I ask two questions:

1. What books/podcasts have you read in the last six months? What did you learn?
2. How do you spend time developing yourself?

Did you know 80% of books are read by 5% of the population? Half of book sales fall in the romance category! Focusing on candidates that are still reading and growing AFTER college has elevated the standards of my team and my hiring pool to the top 2.5%. I also want to see how they spend their time. Candidates who grow themselves are rare and are likely coachable. They value growth and knowledge, and my job becomes ten times easier.

Another way I look at coachability is how a person receives constructive feedback. In my interviews I'll share 1 or 2 concerns I may have. Some candidates love the feedback while others seem stunned that I was so straight with them. If they are receptive, then that's a good sign that they are coachable.

Fire in the Belly (Capability)

What is the difference between employees who move up versus the rest? It's the desire to win! The fire in the belly is needed to continue to learn new technology and keep up with industry news and trends. It's the hunger to work the extra hours per day. It's learning how to be more efficient. It's devoting time on weekends to prep for the week. It's the desire to make the impossible happen to get the deals in ahead of the competition.

"How hungry are you?" This is why a lot of technology hiring descriptions have terms like "self-starter," "hungry," and "proactive attitude" in them. Top performers have the desire to succeed. Top performers also like being measured. Just look at the back of baseball cards—it's all about the stats!

I'm not saying that you have to be athletic to be great with technology. I'm saying there are internal and external motivations, and athletes are a great example of people that have both. It's important to have both: internal drive (the need to be the best, the hunger to make an impact) and external motivation (competition among peers, family to support, nonprofits to help) to drive yourself further than ever imagined.

How do you show you have fire in your belly? You can tell a compelling story. You can take extra actions, above and beyond the regular interview expectations as well. You can take the extra effort to leave a powerful, positive impression.

Intelligence (Capability+Likability)

For most hiring managers your GPA really doesn't matter. How you learn does. Let me put it this way, when you graduate college, can you explain how Salesforce or cloud computing works? Likely not. It's the skill of taking a subject in which you have no prior subject expertise, like Chemistry 101, and being able to articulate it. It's your capability to learn and articulate that I'm referring to when I talk about intelligence.

I look for two things when it comes to intelligence:

» Emotional quotient (EQ): also known as street smarts and thinking on your feet

» Intelligence quotient (IQ): book smarts and being able to study, analyze, regurgitate, and apply

EQ is determined on the phone and onsite interviews. Interviewers for customer-facing roles want to see if the candidate is likable, capable at building rapport, whether they take an interest in the interviewer and focus on the interviewer's needs, how the candidate handles objections, and whether the candidate is a culture fit.

How do most hiring managers test IQ without looking at a GPA? Easy! Tough questions. Interviewers may create mock scenarios that make you think. The most common method I've seen is to ask a sales candidate to do a mock cold call and/or explaining their software either casually or through a mock presentation. For developers the most common is a code test to screen out their actual coding skills. For marketing it could be to create a mock marketing campaign.

After making it past the first onsite interview with the VP of sales, the candidate's next task may be to take three days to prepare for an onsite presentation with the team. The email sent with the instructions and materials will be standardized across all candidates to see how technically deep each candidate can get on their own. I get to see how "smart" they are when it comes to learning a technology solution from scratch and how they articulate it.

Humility (Likability)

The final and most overlooked ingredient in what makes an amazing hire is humility. If you have the four previous qualities, you are likely in the top 2.5% of hires and a top performer. And when you are the #1 employee or rep, it might get to your head. Most were not taught that it is more amenable to be humble and successful than to be openly proud when you are the best. Success without humility will only take you so far.

I personally experienced this during my first year in technology sales. When I joined, I was super humble. I would ask for time with the senior SDRs to listen to their calls and learn best practices. I befriended Steve who I thought was the most tech savvy person on the sales team and sponged sooooo much from him. After learning it was time to execute. Over the first quarter I was head-down trying to figure it out. After about a month of calling and emailing customers, it clicked. I was able to go from booking one meeting per week to 1–2 meetings per day consistently.

The majority of my heightened performance overflowed to the second quarter and—boom—that quarter I became the top SDR in the region and #2 in the company. I didn't think twice about it as I was the new guy. I just kept my head down. It wasn't until I was recognized by my worldwide VP and I was beating out the senior SDRs who trained me that it started to get to my head. It wasn't that I was overly outwardly cocky. It was I stopped being the humble sales rep that everyone knew and loved. I became competitive and wasn't afraid express it.

I didn't even know I stopped being humble until I was asked to have a meeting with my inside sales manager and director. That's when they acknowledged my increase in performance as well as my shift into competition without humility. To this day I'm thankful for their coaching as it's hard to see that blind spot in the midst of competing and being asked to fly out and share your best practices with other sales teams.

It wasn't until I became a hiring manager that I made sure that humility was a high value. We wanted competition and drive, but we didn't want a bunch of sales knuckleheads. I've noticed that all executives I've worked for are super successful and talented, yet very humble. I've noticed that humility and talent make a powerful combination.

When you are young or new to a high-growth environment, there will be blind spots. Just be humble. If you are blessed, seek out mentors who are willing to help you grow and become the best version of yourself.

Rankings

Give yourself an honest grade on where you are today. How would you rank yourself on a scale from 1 to 10 in each of these 5 areas?

1. *Integrity*: do you show up on time? Do you follow through with what you say you are going to do and take responsibility for your actions? Do you communicate with others when you must change your plans? Do you acknowledge it when you make mistakes?
2. *Coachability*: how do you respond to constructive feedback? What's the last software, sales, or self-help related knowledge that you have exposed yourself to for growth?
3. *Fire in the Belly*: how driven are you? What are your external and internal drivers?
4. *Intelligence*: If you had to explain a new software solution like Google's new enterprise Cloud solution, how would you gather info to educate yourself about it? How quickly can you pick up a subject like endpoint cybersecurity?
5. *Humility*: would your family or friends say you're a humble person? Can you describe yourself and your past achievements without sounding like you're bragging?

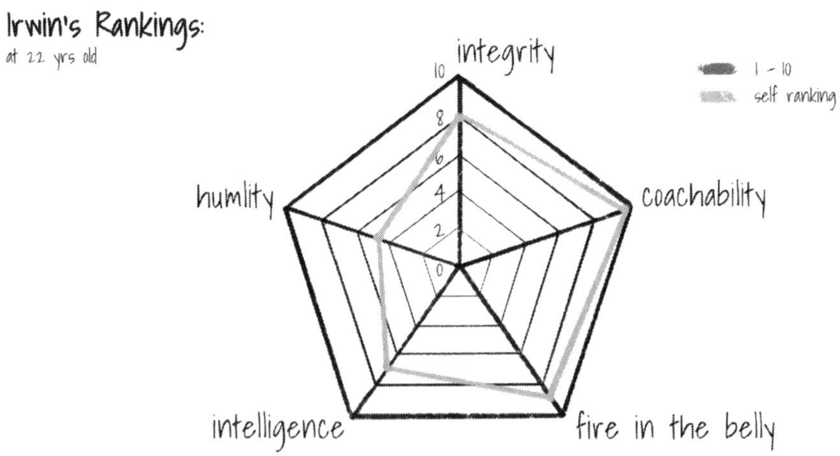

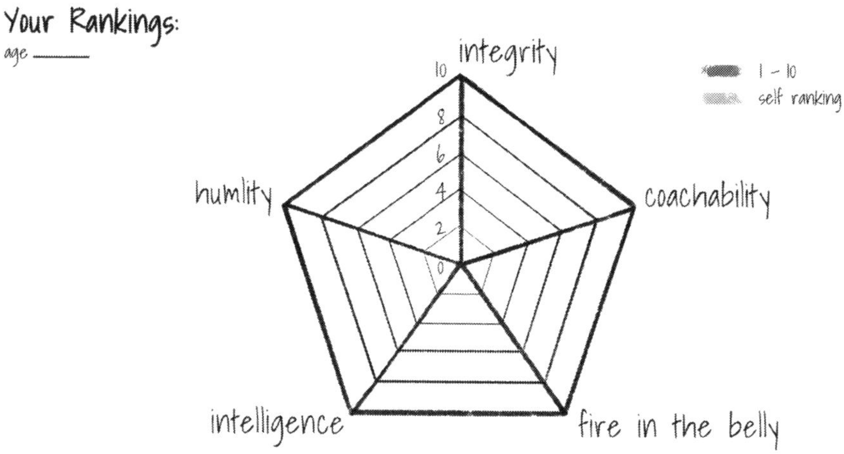

I want you to shade in this pentagon where you are in each of these five areas today. If there are areas you are weak in, then create a plan on how to improve. Improving yourself won't just enable you to deliver a more effective interview; it make you a better team member and a better human being. It's a win-win-win!

Not Convinced

During the interview, here are the common reasons why a candidate doesn't convince me on why they would make a good fit:

1. There is not a clear picture of who you want to be. I'm not sold on your three whys.
2. You are not understanding and addressing my needs as a hiring manager.
3. You don't have presence, energy, and enthusiasm.
4. You are not a culture fit.
5. There's a lack of follow up or you didn't try to "close the opportunity" meaning you don't make the effort to see where you stand, whether you are a good fit, and what the next steps are to move forward.

The best way to help address these pitfalls is to do research, do a good job actively listening to what your hiring manager and teammates are looking for, and be proactive in addressing it. Remember "Losers react, winners anticipate."

At the end of the interview, the interviewer will likely ask whether you have any questions. YES! Always have one good question to ask. There is a difference between good and bad questions. Good questions make the interviewer think and lets them know that you've really thought out the question. Bad questions are questions that require no thought and are self-centered.

Here are examples of bad questions to ask in the first interview:

- How much time off do I get?
- Can we date employees?
- What's the work-from-home or dog-at-work policy?
- What's the salary and benefits package (Save this when you get an offer)?
- How soon do you promote employees?

It's best to ask these kinds of questions later, when you are 90% sure you'll get the offer (usually when you are discussing salary and start date).

Good questions are questions that help show your intelligence, interests, and thoughtfulness. The questions mentioned above are usually around what "I want," and though your needs are important, timing is everything!

Here are examples of good questions:

- » I've noticed you had an amazing career or interest (e.g., hiking Mt. Kilimanjaro). Can you share with me the most important thing you learned doing that?
- » Could you share the current and future growth rate of the company?
- » What are the key initiatives or challenges your ideal candidate could help with?
- » Is there a question I should've asked that I haven't?

If you feel the interview is NOT going well. As in you know in your in your gut that they are likely going to pass on you. I highly recommend taking this opportunity as a learning moment. I would recommend asking, "Are there areas or gaps that I can improve on based on our conversation today?"

GOLDEN NUGGET

One of the best ways to increase your self awareness is to give the people closest to you permission to give you honest feedback on areas they think you could get better in. Then listen with an open mind. If you can avoid being defensive there might be a few things you can learn about yourself.

This question will help you find your blind spots. You'll get to find out what concern or unanswered question is left for the hiring manager. You simply don't know what you don't know. But asking this will give you a chance to help address their concerns. Remember, if you can reduce the perceived risk in their hiring you, then your chances to move forward go up! The sooner you find out, the better you'll be for the next interview.

> **One more thing:** Always send a thank you follow-up. This act that takes a few minutes is worth it. It's important to follow up as soon as you can. It does leave a better impression.

The goal of this chapter is to give you the qualities of an A player and more importantly where you stand today. If you do serious self-reflection, you can prioritize new ways to show that you have the qualities that your hiring manager is looking for. My goal here is to also share with you the common reasons why you might not move forward in the interview process. You need to know the formula for success, so you can crush your interview, and you need to know your red flags, so you can avoid them.

When you become the ideal candidate, you'll get multiple interviews. By having a lot of interviews, the tables turn in your favor. If a company wants you, they will have to compete for you. Your candidate value has increased. In return, you'll have multiple offers. The next chapter aims to help you get the most out of your desired offer.

CHAPTER 19

MASTERING YOUR FIRST OFFER

Have you ever wondered if you should have negotiated or pushed back when you landed your first job offer? It is the top topic people asked me to address in this book.

It's crazy to think that one extra phone call or email could result in tens of thousands of dollars more (sometimes hundreds of thousands of dollars due to equity!).

The very best way to immediately maximize your offer is to have **multiple conversations** with multiple companies at the same time. This favors you in several ways:

» It will increase your perceived value to all companies.
» It will accelerate your interview process.

» It will give you an idea of what the market is offering you.
» It will give you confidence when pushing back on the initial offer.
» It will give you options to make the best choice.

Please notice I deliberately mentioned "conversations," not "offers" or "interviews." Given that when you line up interviews with multiple companies, the stages in the many interview processes are going to vary differently. There are different stages: from simple conversations with friends about a new opening, to actual phone screens and onsite interviews.

Leah and the Power of Multiple Interviews

As already mentioned, I've been mentoring Leah, a server at a high-end restaurant in San Francisco, on how to break into tech. Every week we would work the entire process, beginning with the gateways. After learning about the various gateways, she chose the SDR gateway. Once she decided this, we then got organized by chartering all the actions, dates, etc., in her MAP. After that, we worked on her resume and LinkedIn profile. She built out her target list, focusing on enterprise stage companies. However, when she contacted them she didn't get any responses. Next, she focused on startups, and that's when she got her first interview. She was so excited about it that she did not pursue the new recruiters that were sending her messages on LinkedIn. (All of this I described already in a previous chapter, but to illustrate my point on the power of multiple interviews, please bear with me. We're getting there!)

When she met with me to help her prepare for her only interview, Leah explained that she wanted to put 100% focus on that one interview. That's when I explained to her the importance of having multiple conversations/ interviews and how it would increase her perceived value to the companies and push them to take massive action on hiring her.

As fate would have it, the hiring manager rescheduled the interview. Because Leah wanted the opportunity so much, she quickly wrote to the hiring manager giving the days and times next week she was available. She anxiously waited for his response. As she waited, the minutes turned to hours, and she was freaking out.

That is when I explained to her the importance of building a pipeline of companies to prevent these kinds of freakouts and how the goal was to get her the best opportunity in the 90-day period. If she was focused on only one interview, she could be setting herself up for failure.

Once we focused on multiple opportunities, she was able to line up additional interviews at additional companies. In doing so, her anxiety was drastically reduced. Plus, she was able to have interviews that gave her good practice before having the interview at her top-choice company.

When I was career hunting right after college, I had five different interviews and ended up landing four offers, which led to my career at BMC. When I was focusing on my second career, I had five different onsite interviews that led to two offers, which led me to my career at OpenDNS. This is the name of the game folks. You need to be really good at creating multiple opportunities in order to have the choice to pick the best one.

A Comparison Scenario

Take a look at this scenario where there are two candidates, Bobby and Bobbi. Let's assume they both have zero-tech backgrounds and are applying for the same gateway position.

Bobby has two years of non-tech job experience and is in the middle of the interview process. He has only this opportunity.

Bobbi has the same two years of non-tech job experience. At the end of the onsite interview, she asks what the next steps are in the interview process. After the interviewer responds, "You should hear back next week," Bobbi mentions, "Okay. No problem! I also want to let you know that I'm talking to Twilio, Box.com, and NerdTech, a seed-stage startup. Box.com wants me to make a decision on their offer by the end of the month."

When Bobbi shares this, there's a chance that the interviewer is going to ask themselves, "Why do these companies want her? What do they see in her that I'm not seeing?" It's very likely that the interviewer will dig deeper to see if Bobbi has something special. If they see it, then they may ask her,

"Where are you in the current process and have they given you an offer?"

If the team does decide that Bobbi is a good fit, then when it's time to line up the hiring manager's boss or next person in the process, the interviewer (or a good recruiter) will inform that manager of the multiple companies that Bobbi is talking to and the timeline of her offers. Bobbi's next interview will likely get set up at the earliest time to ensure no time is wasted.

As far as Bobby goes, since he never mentioned he's looking at other places or any offers he's gotten, there's no sense of urgency or piqued interest around hiring him, or at least not to the extent as exists around Bobbi.

Can you see how conversing with multiple companies can help increase your perceived value to a company and accelerate your interview process?

Preparation Is Golden

Compensation will eventually be brought up! Knowing what has been offered before for the position, so you can go in knowing what you want and could get is key to getting the most out of an offer. It will be extremely beneficial to have already researched compensation on Glassdoor.com. By doing this compensation research beforehand, you are prepared when you reach the offer stage. When you are asked, "What is your desired salary range and expectation?" you have a response ready.

As most people reading this book may be coming out of a lower-paying role or industry, be careful to avoid "anchoring" your salary expectation. Anchoring means the first number that is shared that creates the first reference point in a negotiation. A lot of times it locks you in on a number, especially one that is introduced too early or that is too low.

DON'T BE AFRAID OF CHANGE.
YOU MAY END UP LOSING SOMETHING GOOD, BUT YOU WILL PROBABLY END UP GAINING SOMETHING BETTER.

— KUSHANDWIZDOM

In negotiations there are a few rules in getting the best offer possible:

1. Don't be the first to give the first set of numbers.
2. If you do have to go first, the rule of thumb is to go very high if you are a seller (interviewee) or low if you are a buyer (interviewer).

Most readers will not already be earning a higher salary than what they could earn in a tech company. Consequently, let the company offer their package first. By doing your research and having multiple conversations, you'll have a good idea what the market is offering.

Caveat: when making a counteroffer on the first offer, it's very important to do self-reflecting about where you stand in the interview process. If you are getting an offer, that is very good sign. There are very few instances (1 in 50 I know of) that they will withdraw the offer if you push back on it with a counteroffer. Be careful to see if there is any possibility that if you push back, you may lose the offer. If the company extended you an offer, you are likely in a very good spot. I am including this disclaimer in case for some reason you push back and then you lose your offer. It's rare because a lot of times if the company can't give a better offer, they will tell you, instead of pulling the offer. It's important to remember that negotiation is part of the process.

With that said, for you to maximize your offer, you must first understand the levers:

- Signing bonus and/or moving bonus
- Total compensation (commissions and quotas for sales-related roles)
- Equity (Stock options, restricted stock options, profit sharing, etc)
- Start date

» Vacation days
» Work from home policy

When I was a hiring manager, there was only one candidate (out of 20) that ever pushed back on the offer we were proposing. He didn't ask for a signing bonus or an increase in total compensation; he wanted more stock options. I ran the request to my VP of Sales who then brought the request to the co-founder. Our company countered back and provided an increase of 50% in the amount of stock options. To my surprise, he turned down the opportunity and pursued work at another startup.

Why was he the only one that pushed back? He told me he spoke to his mentors who told him to ask for more equity.

Why don't more people counter the original offer? Most folks out of college or new in the industry are excited and grateful to get an offer, so they accept it without considering whether they could have gotten more. That was exactly my mindset when I first got my job offer at BMC. I had two other offers, and the BMC one was the highest and my #1 choice, so I accepted it without pushing back on any of the levers previously mentioned. Afterward, I learned that only one of my other friends pushed back and he ended up getting a small bump in his salary. Looking back, I learned that doing a counter offer is normal and not frowned upon. They'll simply say it's the best they can do.

Honey over Vinegar: Negotiation Lesson with Shane

Remember Shane and his seven offers (chapter 17)? Like most folks starting out, he had no idea how to approach negotiation. He didn't even know if he should or could. When I taught him the five levers, we determined the most important was crafting a REASON as to why he was asking for more. Let's recap what offers he was able to drum up!

» Offer 1: $65k compensation only (no bonus, equity, etc.)
» Offer 2: $70k compensation only (no bonus, equity, etc.)
» Offer 3: $75 compensation only (no bonus, equity, etc.)

When we first did a roleplay, and I pretended to be the hiring manager, Shane fumbled through his hellos and was too direct in his asks! "I would like a higher salary, a signing bonus, and equity." Trying to negotiate like this is awkward and likely won't work.

When we switched roles, and I played the "negotiating candidate," it played out like this:

Negotiating Candidate [played by Irwin]: Hi, Shane, how are you doing?

Hiring Manager [played by Shane]: Doing great. How are you and how can I help you?

Negotiating Candidate: Really good, actually. I've had a lot of good traction with the job search. I was able to land three offers, which is why I wanted to hop on the phone with you today. I'm entertaining and I'm really excited about your company. I spent a lot of my own resources moving out here, and I was wondering if you guys offered a signing or moving bonus to help.

[*Wait for response. After response, continue ...*]

Negotiating Candidate: My top offer was for a $75k compensation, and I wanted to see if there was flexibility on in matching it. They are from my number two choice, and I would love your feedback if you guys can be competitive.

[*Wait for response. After the response, continue ...*]

Negotiating Candidate: And finally I wanted to get your feedback on equity. Part of the reason I moved out here was to join a startup and take part in building a company with you and the founders. Is there a reason why equity was not discussed?

[*Wait for response.*]

After Shane heard my approach, a light bulb went off. He realized that negotiations are more of a conversation than a demand. Starting with a warm opening and humble bragging that he has many suitors helps shift the power in his favor a bit more. He also learned that for each lever he is asking for, he must craft a reason (or story) as to why he's asking for it. Finally, the smoothest way to ask for something is to simply ask a question versus demanding what he needs. It's all about the delivery, and by making it more of a conversation, it helps make the process flow smoother for both parties.

*

When I was negotiating with my 1st tech startup, I understood the offer game a bit more. Plus, I already had a promotion with a compensation plan that was the same as what they were offering. I requested they help cover the cost of my moving my life across the country and that they beat my current compensation plan as San Francisco was more expensive for me to live in. I asked for more stock options and that they extend the start date, so I could enjoy a month with my girlfriend and get my life settled in San Francisco.

Outcome: They were able to pay for my move, increase my salary by $10,000 per year, and increase the amount of my stock options. I was very glad I asked for more.

When I was interviewing for a role at my 2nd startup, I also had another offer from another startup and their total compensation plan was $20,000 more per year. So I explained this to my hiring manager that the competing startup was offering more and I was leaning towards that opportunity. My hesitation worked in my favor as the VP of sales jumped in the interview process and gave me a call. At the time of the call, I'd already made the decision in my mind to move forward with the competing startup. I explained the situation to him and asked for more on all four levers mentioned earlier.

Here was his counter offer:

1. He offered a $5,000 signing bonus
2. He explained that the salary+commission was the max compensation he was able to get approved.
3. He offered that during the first 2 months he would pay my full commission in my contract. As sales reps don't get paid commission until they close business which could be 3-6 months from the start date.

4. The shares were fixed, and it was the best he could do. He did remind me that the shares dilution was very low (and it definitely was; the number of total shares was less than the number of shares issued at the competing startup), which means that in effect, they were offering more equity.
5. We mutually agreed on a start date as he wanted me to start as soon as possible.

In this way, he did narrow the delta. At the same time, having the other offer helped me get the most out of my offer. The biggest takeaway is to prioritize what is important and if you have the courage ask for all the levers possible. The worst is what you saw above where he just simply said it was the best he could do.

This isn't just my scenario. I've tried it multiple times and helped my family and friends negotiate their offers. Every time we have countered with the chosen offer levers. We were able to get anywhere from $5,000 to $20,000 more out of the final offer!!! Which is crazy and life changing for someone starting out their first career! The number could be in the six or seven figures if you were able to get more equity out of a startup that turns into a unicorn or decacorn ($1B or $10B valuation).

Key Takeaways

» Know where you stand. Research Glassdoor.com to get inside information on interview processes and average compensation for your roles. If you have a friend in the company having that insider information helps as well.

- Understand that pushing back (countering) is part of the interview process, and 99% of the time they will NOT withdraw it. They'll likely let you know what they can and can't do.
- Having multiple conversations with companies will help increase your candidate value.
- When asking what the next steps are, mention the companies you are speaking with and any offers and interviews you already have on the table. This will accelerate your next steps.
- Understand the levers in an offer and how you ask (inquisitive, friendly, etc) is important!
- Having a higher offer or late-stage conversation with another company will give you leverage and confidence. Go for more than one company.
- Have fun! This process can be very enjoyable and fun. At the very least, don't forget that you are learning something in the process!

I hope this chapter demystifies how to get the most out of your offer. If you are here, then awesome. Congrats on getting your first offer! It's a huge step towards your goals. The good news is you are so close to finishing this book! In the next chapter I'm going to tie the concepts together for you!

CHAPTER 20

TYING IT ALL TOGETHER

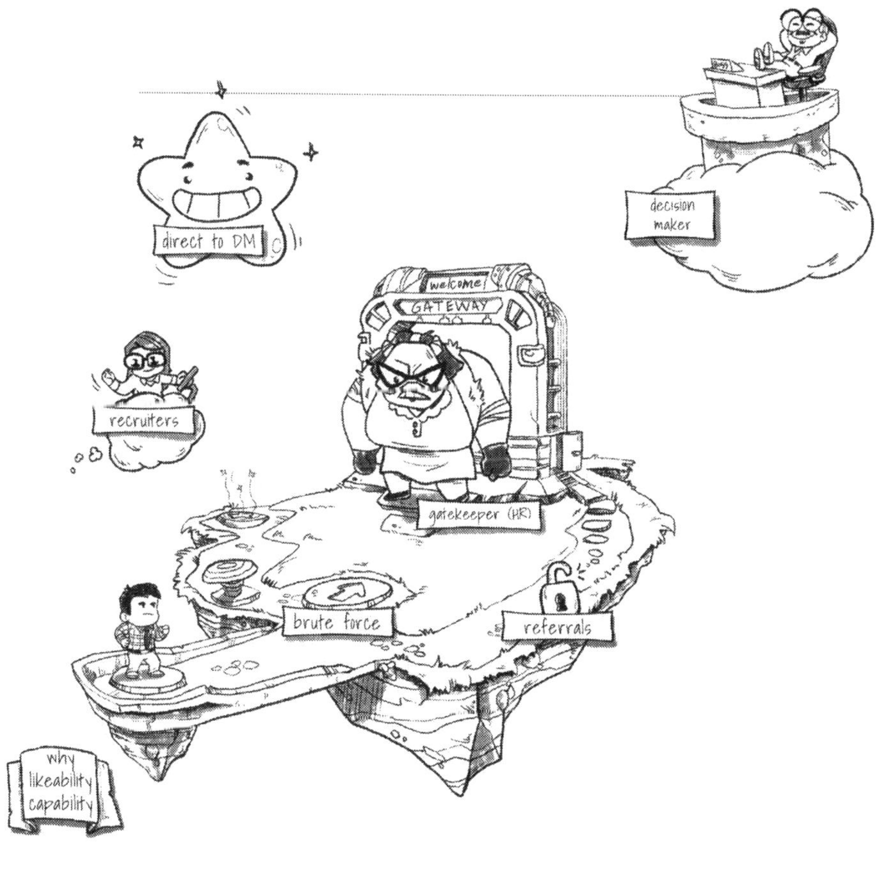

Whether you are a college senior looking to make the next step in life or just looking to break into a new career, *Breaking into YOUR Silicon Valley* has laid out all those next steps for you.

If I had to summarize the lessons this book teaches, particularly that part 3 lays out, here's what it would look like:

SEEK CLARITY!

You need first to find your gateway and figure out where you want to work and who the ultimate decision-makers are.

EXECUTE STRATEGICALLY!

You want to apply like crazy (like Shane did) through brute force or strategically find internal referrals and recruiters and/or reach out to the decision-makers directly. All the while, you must define your WHY and keep it top of mind! Then focus on your likeability and communicate your capability to make the difference you are looking for. This is the framework on how to break into any company and create the opportunities you are seeking.

YOU HAVE CHOICES!

You've learned too that tech careers aren't solely about coding, math, and computer science. In fact, there are several gateways for breaking into tech. Entering tech companies through a gateway role is a tried-and-true system that will allow you to take the first step to achieving your dream job. Gateway roles are how you bridge the gap between college and the "real world" of tech.

SUCCESS LEAVES CLUES!

You've read interviews from real people who have used gateways themselves. Many of the interviewees were self-taught, had no tech backgrounds, and stumbled into tech careers that they are now flourishing in. They've shared their paths in detail, so you can benefit from their stories to gain inspiration, motivation, and more options.

UNDERSTAND THE PROCESS!

You've been given the necessary foundational recs, insider tips, and advice, so that you can craft your resume in a way that shows you are the right person for the tech job, even if you have no previous tech experience. You've even been supplied the mutual action plan (MAP) to help you layout and organize every step along the way to make sure you take massive action to achieve your dream. You've been provided recommendations on how to build your target list, reach out, master multiple interviews, and negotiate multiple offers.

Now, it's up to you: choose your tech gateway, lay out your MAP, and make the first step towards your dream career.

My goal is to ensure that the next generation gets an inside peek on the gateway jobs that can lead into the tech careers of tomorrow. Transparency is a value I believe in as do the companies I've worked for. My hope with this book is to give you important insight into the reality of tech career possibilities, so that you have the information, courage, and inspiration to join the exciting world of high-tech.

So what could possibly stop you from achieving your dreams?

I'll give you a hint:

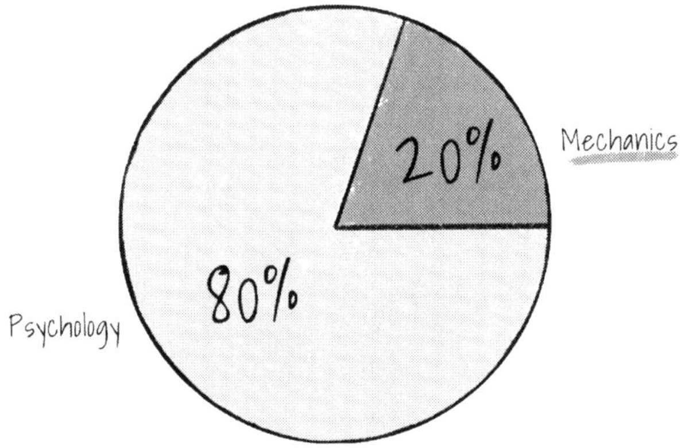

The only thing that will stop you is your psychology (aka mindset & limiting beliefs)!!!

The part 1 of the book was designed to help address limiting beliefs, concerns, doubts and teach you about the benefits of working in the tech sector and how a gateway job is different than a entry level job. Part 2 was designed to educate on the different options available as well as strengthen your beliefs by hearing different stories from various people who had unconventional ways to breaking into the tech industry. Part 3 was designed to lay out the step by step strategies from blank resume to negotiating an offer.

What isn't in this book is the mindset exercises and psychological training I've learned to help those who are still struggling. If you are still stuck from taking action and you would like additional help. I am planning to build additional content for those who need it. If that's you...go to www.IrwinKi.com and add your email to any of the email opt-in fields and shoot me a note that you would like additional training.

If you found my book valuable, please share it with someone else. Also check out **www.IrwinKi.com**, as I'll be posting thoughts, ideas, and interviews on the subject of breaking into tech.

Warm Regards,
Irwin

PS

Cryptocurrencies are a hot industry, and there are a lot of opportunities there. Learn more about it in the first bonus chapter!

When I was growing up, I spent a lot of time playing video games. I bet a lot of people reading this would also like to know more about how to break into the gaming industry. You'll find a second bonus chapter dedicated to this subject.

Check out these two bonus chapters and happy career hunting!

THE ROAD TO SUCCESS IS
NOT EASY TO NAVIGATE BUT WITH HARD
WORK, DRIVE, AND PASSION, IT IS POSSIBLE
TO ACHIEVE THE AMERICAN DREAM.

— TOMMY HILFIGER

GET YOUR FREE BONUS!

Thank you for reading & sharing this book!

To help with your **momentum!** Here are 4 **FREE videos**

I've included for you to bring the **book to LIFE!**

Go To: www.irwinki.com/p/FREETRAINING

If you want to see more videos!

Follow me: Search "Irwin Ki" on Youtube

Find me via @Irwin_Ki on Instagram

Get **connected with other readers** via FB group:

"Breaking Into Your Silicon Valley Community"

BONUS

BREAKING INTO CRYPTO

Cryptocurrencies are going gangbusters! There was more money raised in 2017 via initial coin offerings (ICOs) than in all of venture capital combined. Silicon Valley veterans are seeing similarities between this cryptocurrency extravaganza and the dot-com boom in the 2000s. I've met 25-year-old multi-millionaires, and I've heard of stories of people who worked for some of these cryptocurrency startups that were paid in cryptocoins and now they are paper millionaires.

If you work for crypto-exchanges, like Coinbase, you could have a percentage of your salary paid in bitcoin on top of your equity. For the employees who did enroll in having 10% of their salary put into digital currency, they could have easily tripled or quadrupled their year earnings depending on when they joined.

The flipside—along with these fabulous opportunities comes risk. A lot of these crypto startups will likely go to zero, but we can all agree that blockchain is not going anywhere. If you don't know exactly what blockchain means, youtube or google how blockchain works for an easy-to-understand explanation.

This, then, begs the question—if it's like the dot-com bubble, how can a person capitalize on it (before the bubble bursts)?

- Become an investor—knowing when to get out is key.
- Become an entrepreneur—create a coin or startup, and add value!
- Service these entrepreneurs and investors. Think California Gold Rush—the people who made the money were the ones selling the shovels. Simply put, the only corporation still in existence after the Gold Rush is Levi's Jeans. Being in the services business (ICO marketing, security, legal, etc.) can result in making a ton of money from cryptocurrency startups.
- Work at the next big crypto startup and get equity and/or tokens with your compensation package and diversify your holdings.

After attending multiple conferences, I interviewed many crypto-entrepreneurs and investors and asked the same questions I asked my other colleagues. What is the best way to break into crypto? Here is a summary of these conversations:

- Start researching cryptocurrency.
- Get in the game! Start with a very small position into bitcoin and build an understanding of how it works.

- Go to crypto conferences where you will immerse yourself in the community and learn from industry experts, investors, and entrepreneurs. You'll also rub shoulders with people who are servicing these companies.
- There are gateway jobs into crypto that are similar to the ones mentioned in this book. Coinbase.com announced they were hiring 200+ more support roles for their company because their current team couldn't handle the number of tickets that were being created as 2017 was a gangbuster year.
- A unique crypto gateway role is a "community manager." This is the person who manages a crypto startup's social media channels: Telegram, Twitter, Youtube, etc. It's important to have an online presence as there are a lot of scams out there, and by being transparent you can raise more money.
- If you are a developer, learning blockchain WILL increase your value. You can command a high pay in this hot space.

If you ride waves (aka trends), the kinetic energy of the wave will help you move forward faster. The bigger the wave, the faster you'll go. Blockchain is a HUGE wave that's happening.

The key is getting in the game! There are going to be plenty of startups in this space, and what chapter 16 (about building your target list) covers about how to choose a winning startup applies here. Remember—follow the smart money. Following the venture capital and entrepreneurs is key here.

I hope you enjoyed this crypto bonus content—and there's more to come! Up next is a passion chapter for those who are interested in breaking into the video game industry. Enjoy!

BONUS

BREAKING INTO THE GAMING INDUSTRY

This book would not be complete without mentioning one of my childhood passions: VIDEO GAMES. It's worth noting that most gaming companies will not offer equity, so that is an important difference from the technology space. Take that into consideration when choosing a career. Because gaming is such a passion type of industry, there is a lot of talent naturally flooding that way, so the compensation is usually less than technology, and the competition is likely more intense.

That being said, I know that a lot of people LOVE video games. And, like all industries, there are gateways.

Gateways

- » Assistant product manager ($74k–$150k)
- » Video game tester ($8-20/hr)
- » Localization specialist ($31k–$88k)
- » Twitch streamer ($0–tens of thousands to hundreds of thousands, depending on skill)

Career Paths:

Product manager, level designer, game designer, localization manager, Twitch superstar: look up "NINJA or pewdiepie" on Twitch and Youtube

Skills Needed

- » Love for video games
- » Project management and communications skill to be a PM
- » Knowledge of foreign language and culture to be a localization specialist
- » Humor and/or video game skills

Justin's Take on Breaking into Gaming

I want to introduce you to Justin who works on the computer graphics (CG) side of Square Enix, the publisher of the Final Fantasy series. Justin was super helpful to bring to light a couple of gateways to break into in the ever-growing video game industry.

Justin: From the time I was young, I wanted to be a game designer. When I first started out, however, I didn't know how I was going to break in. So, I started studying computer graphics (CG) and 3D animation. After some time in the CG industry, I moved into game development. I taught a lot of it to myself.

There's few ways you can get in. You can actually teach yourself some coding, or you can learn applications. For example, these days you have Unreal Engine and Unity, which is just free game development software. You can dive into it and learn. And you can create your own game. So, that's the path I used to get into where I am now.

Irwin: Did you study in school for that?

Justin: Yes, I started in a CG school, but it was mainly on 3D animations, which is visual effects, animation, Pixar, Dreamworks, animation movies type stuff.

Irwin: What do you love most about what you do?

Justin: Most people think that to get into computer graphics games, you need to be really creative. But if you're more logic-oriented, more on programming, you can pursue games as well. I'm not a very creative person, but I'm a logical thinker. So, I went into the development section of animation. And that's what led me to game development.

Irwin: What are the different ways in which a non-programmer can break into the gaming industry?

Justin: If you do not have any background in coding or programming, you can always start as a game beta tester. These days there are so many multiplayer online games. Be a beta game tester, and grow from the way you dissect a game to learn how to debug it. Then, send your reports to the game company. That actually serves as experience for you to get into a proper game beta testing position.

Irwin: From being a game tester, where can you go in your career?

Justin: First, you can be part of the gameplay programmer. You could be a level designer, or ultimately you could become a game designer. You can slowly climb the ladder and get into the main lead game designer position. That would be one of top things you'd want to do, as someone who is really interested in game development.

Irwin: What kind of qualities makes a good game tester? What skills or what type of person would do well as a game tester?

Justin: As a game beta tester, there are two ways you can get in. One, you could be someone who follows the storyline and the gameplay, evaluating how it goes and develops as an experience to you. Based on that experience, you learn how a game was made to make the player feel connected into it.

Another way is, you would be a troublemaker. When a game asks you to go left, you'd go right. When it asks you to proceed with story, you just go back. You try to break the expectations of the game to see what bugs there are in it. Next, you learn how to report that in a clean, nicely written form, and you submit that information to the company that made the game. That's actually really helpful for the game company because in games that have open beta, that's the whole point. They want people to test out the game and tell them what the bugs are in it.

Irwin: What do you think makes a good tester become a game designer?
Justin: To make that jump, it's definitely based on a lot of experience. Try out all kinds of games, without any bias to the type of games. As a game

tester and a game designer, you don't want to be biased to certain types of games. You want to try every game. Be interested in every game.

If you really want to get into a game designer position, try to create your own game. That's the best experience you can give yourself—creating a game. In doing so, you give yourself the real experience of what you need to look for and what the user is experiencing. You have to answer these questions for yourself: what do you want to convey to the player? What do you want the player to feel? What kind of story actually interests the current market? All of this is part of the path to becoming a good game designer.

Irwin: What are the other ways in which you can break into the gaming industry?

Justin: The two that I can think of are project management and localization specialist. About project management: in order to create a game, movie, or anything, you need to be able to work with a lot of "actors," like the resources that you have, the people that you have, the budget that you have, and the time that you have. The project management person actually oversees the whole thing.

Irwin: How does a person become a project manager? What are the qualities that make a good candidate for a project manager?

Justin: You really need to connect with your team. For me, that's number one, based on my experience. Really understand your artists, their capabilities, and their feelings.

And from there, the project manager works with them on a given budget. Budgets can't really change very much. So, how fast can you think of solutions when problems happen? Because when you're dealing with people, things happen. You're dealing with art. People view art so differently, every single individual. So, you need to be able to listen and understand what the director wants, what the producers want, and what your artists can do, want, and feel. It's quite a lot of factors. I think the number one key is to really understand. Then the next key quality is the ability to communicate effectively.

Irwin: Tell us about using localization to break into the gaming industry.

Justin: To explain localization, let me start with an example. The Final Fantasy series was developed in Japan. So, the main language of that is Japanese. But if you want to sell the game in the United States, you need it to be in English. But you can't just translate it word for word. You need to understand the slang and nuances of American English. You need to understand different people from different cultures, different cities, and different backgrounds. They all have their own local words that they use for various phrases that differ from place to place. How an American might say something is different than how an Irish or British person might say it. So, localizing takes all of that in account, on top of translating it into that particular language.

You don't really need to understand games to get into gaming using localization, but you need to be able to understand people's cultures, people's ways of life, and the way people in specific regions phrase things.

Irwin: What's the career path for a project manager?

Justin: For project management, most of the time, you start as an assistant production manager or an assistant project manager. Then, you go up to project manager, senior project manager. And in the long run, you become a producer. Once you really understand the whole of what it takes from A to Z to create a game, you can end up being a producer. And you can do a bit more networking in the industry.

Irwin: What about the career path for localization?

Justin: For localization, you can start off as a basic localizer. And then you slowly ascend the ladder as you direct localization teams. Sometimes you get to work with teams from different countries because you'll need people of that expertise, or of that country's exposure, to be able to give that experience that you want the gamers of that particular country. Then you move up to senior lead, and maybe even regional localization.

Irwin: Do you have any additional advice for new folks breaking into the gaming industry?

Justin: If you want to break into the game industry, you shouldn't be afraid of programming or computer graphics.

*

As in all of my interviews, I had a blast talking to Justin and learning about the different ways to break into the gaming industry. I was pleasantly surprised about the idea of making your start in gaming by creating your own games. It totally makes sense. Creating games, no matter how simple the game is, gives you that particular insight needed to articulate what makes a great game.

When you create a game there are going to be bugs. The video game beta tester is similar to the QA tester in chapter 8 The Developer Gateway, except instead of looking at code, line by line, you are playing a game. As an assistant product manager, you get to learn the ins and outs on how to manage a game to completion.

Hope this bonus chapter gives you a good start if you're interested in breaking into the gaming industry!

ACKNOW-LEDGMENTS

I want to thank—my wife, Marisa, for her patience and love in this process; my parents, Angie and Wilson, for the risks taken to make it to the US; Connie Osterholt, for helping me find the courage to take on the idea of writing this book; Adam Aarons for teaching me to lead from the front and Tony Robbins for showing how impacting lives leads to a passionate and fulfilling life.

To Chelsea Miller, Nancy Pile, Chandler Bolt, and Sean Sumner for being the best writing team! I couldn't have done it without this team!

To Ida Fia Sveningsson for an incredible cover & my man Toby Cochran (He is a Creative Force of nature living in beautiful San Francisco as a Director, Illustrator and Storyteller. You can find him via www.tobycochran.com) who is the artist behind the AMAZING illustrations in this book. The book was not the same without your touch.

This book would not be as detailed without your voices too! Thank you, Courtland Smith, Ross Nichols, Aisha Stephenson, Jason Souw, Angelina Columbo, Luke Lin, Len Boyette, Jim Knutson, Wayne Herman, Josh Pyorre, Fallon Neuvert, Jeff Rachlin, Jeff Samuels, Clint Harris, Ryan Crump and all the folks who shared their time, stories, and wisdom for the next generation.

Thank you, Todd McKinnon, Freddy Kerrest, and David Ulevitch, and the numerous angel and VC investors. You have created an opportunity that has shaped my life, and you've instilled in me memories, stories, and inspiration to pay it forward.

ABOUT THE AUTHOR

Irwin grew up as rebellious adventure-seeker trapped in a small town. He was bit by the sales bug at an early age when he discovered making money by selling candy and lemonade to classmates in elementary school. As a teenager, he discovered one of the secrets to breaking into a highly competitive environment when he applied to his state's most esteemed magnet high school: learn the insider secrets from someone on the inside.

Irwin left North Carolina and moved to Silicon Valley with his wife to pursue technology startups. Since then, he has helped build sales teams, won numerous sales awards, including one for breaking the record for the largest deal, attended multiple president's clubs, and has been a leader and mentor on his team. One of the earlier startups he worked with, OKTA, has gone on to list on the NYSE, and the second, OpenDNS, was acquired by Cisco. Reflecting back, he sees how his joining the tech industry has positively impacted his family, friends, and career. It's been a win win, and the game's still going!

His mission is to help share insider secrets so that regardless of your background, you can start a career in technology and change the world.

To learn more, check out **www.irwinki.com.**

SOURCE

Schonfeld, Erick. “The Top 10 VC Firms, According To InvestorRank.” *TechCrunch*, TechCrunch, 25 May 2011, techcrunch.com/2011/05/25/top-10-vc-firms-investorrank/.

Meisler, Laurie, and Christopher Cannon. “U.S. Venture Capital Swells and Spreads.” *Bloomberg.com*, Bloomberg, 25 Jan. 2018, www.bloomberg.com/graphics/2018-venture-capital-deals/.

“Chapter 10, Get to a City.” *FOUR: the Hidden Dna of Amazon, Apple, Facebook, and Google*, by SCOTT GALLOWAY, PORTFOLIO PENGUIN, 2018, pp. 369–370.

“Chapter 6, What’s So Special About Silicon Valley?” *Angel How to Invest in Technology Startups - Timeless Advice from an Angel Investor Who Turned $100,000 into $100,000,000*. By JASON CALACANIS, Harper Business, 2017, pp. 52-54.

Kedemy, Dan, et al. “The Beauty of a Bad Idea.” performance by Reid Hoffman, et al., season 1, episode 3, 2017.

Bolt, Julie. “25 Enterprise Startups To Bet Your Career On.” Business Insider, 7 Feb. 2013, https://www.businessinsider.com/best-enterprise-startups-for-your-career-2013-2.

Bolt, Julie. “21 Enterprise Startups To Bet Your Career On In 2014.” Business Insider, 9 Jan. 2014, https://www.businessinsider.com/21-enterprise-startups-to-bet-on-2014-1.

“Chapter 2, The Early Years of Uber.” *The Upstarts: How Uber, Airbnb, and the Killer Companies of the New Silicon Valley Are Changing the World*, by BRAD STONE, Little, Brown and Company, 2017, pp. 107-109.

32404263R00227

Made in the USA
San Bernardino, CA
13 April 2019